THIS IS NOT A DRILL

AN
extinction rebellion
HANDBOOK

=> THIS WAY TO SAVE THE PLANET =>

PENGUIN BOOKS

UK | USA | Canada | Ireland | Australia
India | New Zealand | South Africa

Penguin Books is part of the Penguin Random House group of companies
whose addresses can be found at global.penguinrandomhouse.com

First published 2019
007

**Edited by Clare Farrell, Alison Green, Sam Knights
and William Skeaping**

Set in 11.15/14pt Crimson Text
Headlines in FucXed Caps
Typeset by Jouve (UK), Milton Keynes

Printed and bound in Great Britain by Clays Ltd, Elcograf S.p.A.

A CIP catalogue record for this book is available from the British Library

ISBN: 978–0–141–99144–3

CONTENTS

THE EDGE OF A CLIFF

— *Mike Barrett, WWF*

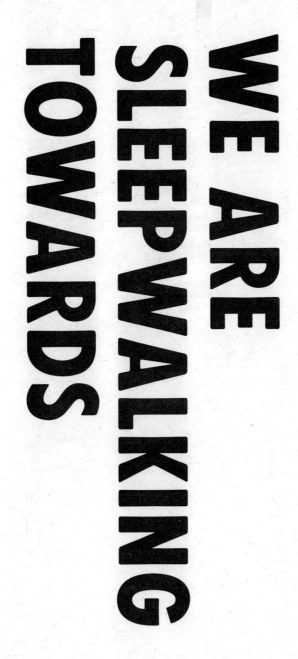

WE ARE SLEEPWALKING TOWARDS

DECLARATION
OF
REBELLION

We hold the following to be true:

This is our darkest hour.

Humanity finds itself embroiled in an event unprecedented in its history, one which, unless immediately addressed, will catapult us further into the destruction of all we hold dear: this nation, its peoples, our ecosystems and the future of generations to come.

The science is clear: we are in the sixth mass extinction event and we will face catastrophe if we do not act swiftly and robustly.

Biodiversity is being annihilated around the world. Our seas are poisoned, acidic and rising. Flooding and desertification will render vast tracts of land uninhabitable and lead to mass migration.

Our air is so toxic the United Kingdom is breaking the law. It harms the unborn while causing tens of thousands to die. The breakdown of our climate has begun. There will be more wildfires, unpredictable super-storms, increasing famine and untold drought as food supplies and fresh water disappear.

The ecological crises that are impacting upon this nation – and on this planet and its wildlife – can no longer be ignored, denied or go unanswered by any beings of sound rational mind, ethical conscience, moral concern or spiritual belief.

In accordance with these values, the virtues of truth and

the weight of scientific evidence, we declare it our duty to act on behalf of the security and well-being of our children, our communities and the future of the planet itself.

We, in alignment with our consciences and our reasoning, declare ourselves in rebellion against our government and the corrupted, inept institutions that threaten our future.

The wilful complicity displayed by our government has shattered meaningful democracy and cast aside the common interest in favour of short-term gain and private profit.

When government and the law fail to provide any assurance of adequate protection of and security for its people's well-being and the nation's future, it becomes the right of citizens to seek redress in order to restore dutiful democracy and to secure the solutions needed to avert catastrophe and protect the future. It becomes not only our right but our sacred duty to rebel.

We hereby declare the bonds of the social contract to be null and void; the government has rendered them invalid by its continuing failure to act appropriately. We call upon every principled and peaceful citizen to rise with us.

We demand to be heard, to apply informed solutions to these ecological crises and to create a national assembly by which to initiate those solutions needed to change our present cataclysmic course.

We refuse to bequeath a dying planet to future generations by failing to act now.

We act in peace, with ferocious love of these lands in our hearts.

We act on behalf of life.

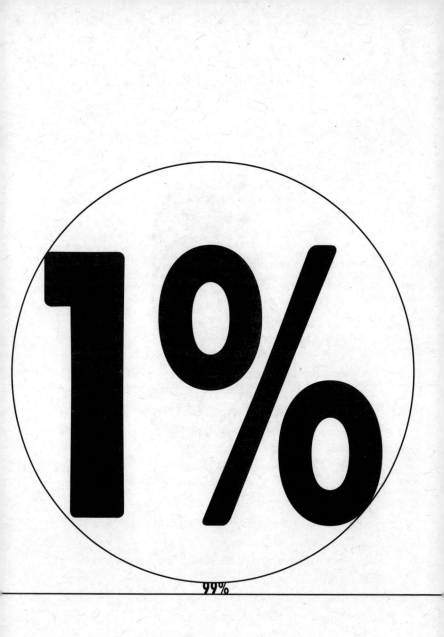

FOREWORD

VANDANA SHIVA

Academic

The signs are loud and clear. From the Earth. From science. From women. From children. From indigenous communities. From our daily lives.

The life on this planet, and our own future, is under severe threat.

We are living through the sixth mass extinction, driven by the limitless greed of the 1 per cent, their blindness to the ecological limits the Earth sets and the limits set by social justice and human rights. We forget that we are one humanity on one planet. There is no planet B. This is where we will live, or go extinct as a species, with the millions that have been driven to extinction by the violence and carelessness of the brute force misleadingly called the economy.

'Economy', like 'ecology', is derived from *oikos* – our home, the Earth. An economy that destroys our home is no longer an economy. It is a war against the planet, the people and our future.

The Hopi people of North America describe the phenomenon of destroying everything that sustains a society as *Powaqqatsi* – 'an entity, a way of life, that consumes the life forces of beings in order to further its own life'.

The *Powaqqatsi* phenomenon of the Hopi is clearly in evidence today. We are dealing with a destructive force that is taking out life forces wherever it can. If the corporations have their way, our fragile web of life will be poisoned and broken, species will be driven to extinction, people will lose all their freedoms to their seed, to their food, to their knowledge and decisions, and all social relations will be ruptured and broken.

Life, society and democracy are under threat. We refuse to allow this future to unfold. We love the Earth; we embrace humanity. We celebrate our biological and cultural diversity and we will defend the rights of the Earth, and the rights of all its citizens, including the last child, with our fearless love and compassionate courage.

To make peace among people we need to make peace with the Earth. To defend the human rights of people we need to recognize the rights of Mother Earth. We need to live through our creativity and her generosity to reduce our ecological footprint while expanding our planetary consciousness of being an Earth family, with one common home.

The extermination of biological diversity and of indigenous cultures that know how to live in peace with Mother Earth is part of one extinction, one interconnected war against life. Ecocide and genocide are one indivisible process, and they began with the idea of the colonization of the Earth as the 'civilizing mission' of a 'superior race'.

In his fifth annual message to Congress on 3 December 1833, US President Andrew Jackson said:

> *That those tribes cannot exist surrounded by our settlements and in continual contact with our citizens is certain. They have neither the intelligence, the industry, the moral habits, nor the desire of improvement which are essential to any favorable change in their condition. Established in the midst of another and a superior race, and without appreciating the causes of their inferiority or seeking to control them, they must necessarily yield to the force of circumstances and ere long disappear.*

And the tribes did disappear. In 1492 the estimated population of indigenous people in the continent north of Mesoamerica was 18 million. By 1890 the Native American population had dropped to 228,000.

Gandhi wrote in his book on freedom, *Hind Swaraj, or Indian Home Rule*, 'This [attacking] civilization is such that one has only to be patient and it will be self-destroyed.' However, in the process of this civilization's self-destruction, it is destroying the planet and our lives. It is destroying our future.

Therefore, it is a moral imperative to rebel against a system that is driving extinction, exterminating species and cultures. To not cooperate has become a moral imperative – a survival imperative. The non-cooperation must begin with the refusal to accept that a system based on ecocide and genocide qualifies to be a 'civilization'. There are better ways to live, to produce and to consume. Extinction Rebellion begins with the liberation of our minds from colonizing categories. We are diverse but equal – not 'superior' or 'inferior'. The extermination of life in its diversity cannot be justified by declaring other species and other cultures 'inferior creatures of God'.

The Earth is for all beings, today and tomorrow.

I call Extinction Rebellion '*Satyagraha* for Life'. *Satyagraha*, for Gandhi, was non-cooperation based on the force of truth.

Today's struggle for truth is that extinction and extermination are not inevitable. They are crimes against the Earth and against humanity. And we can stop this crime by refusing to participate in and cooperate with this project of ecocide and genocide.

Together, as diverse species and diverse cultures, we have the creative power to stop extinction through non-cooperation at every level, beginning with each of us, expanding the rebellion into 'ever-widening, never-ascending circles' of interconnected life and freedom.

This is the call of Earth Democracy. This is our highest duty as Earth citizens.

INTRODUCTION: THE STORY SO FAR

SAM KNIGHTS
Extinction Rebellion

This book is about a rebellion. A rebellion that is happening now.

Although perhaps, by the time these words are published, the rebellion will be over. Perhaps the rebellion will have died, suddenly or without warning. Perhaps it will have simply vanished, consigned to a long history of failed revolutions and fruitless campaigns. Perhaps it will have been successful. Perhaps you will now all be deep in the process of constructing a new kind of world. A world in which all human beings are created equal, and no person, or party, or corporation sits in dominion over their fellow human beings or – indeed – the Earth.

It is impossible to overestimate the significance of where we are now. The journey here has been long and arduous. It has been fought for by thousands of scientists, academics and activists all across the world. It has been hard, and punishing, and sometimes very lonely.

Extinction Rebellion began in a small English town. It began with fifteen people who had studied and researched the way to achieve radical social change.

Together, these fifteen people decided to embark upon a long campaign of civil disobedience, a campaign that would transform the way in which we talk about the climate and ecological emergency and force governments all over the world to act. We started touring the country, visiting communities, villages, towns and cities. We gave talks, took action, and slowly began to build a movement. The talks we gave were clear, straightforward and led by science. We walked people through the facts and then, at the end of the presentation, provided a necessary and rational response: mass civil disobedience.

So, on 31 October 2018, we declared ourselves to be in open rebellion against the UK government. Since then, in such a short space of time, hundreds of Extinction Rebellion groups have been established in countries across the globe. The movement is now active in every single continent except Antarctica. Hundreds of thousands of people have signed up to block roads, shut down bridges and – if need be – to get arrested.

In April 2019 we began our first phase of International Rebellion. In Pakistan, we marched through the capital. In the US, we glued ourselves to a bank. In the Netherlands, we occupied The Hague. In Austria, we blocked roads. In Chile, we lay down in the middle of a street. In Ghana, we blew whistles to sound the climate alarm.

In the UK, meanwhile, we shut down five iconic locations in central London: Oxford Circus, Marble Arch, Waterloo Bridge, Piccadilly Circus and Parliament Square. We stayed there for ten days, delivering a rolling programme of speeches, discussions and public assemblies. We closed down fossil-fuel companies, blocked the roads around the Treasury and glued ourselves to the London Stock Exchange. We attempted to cause as much economic disruption as we possibly could.

By the end of the fortnight over a thousand people had been arrested.

The protests had cost the city tens of millions of pounds and completely confounded an already stretched police force. We were quickly invited to meet senior politicians from all the major parties, including the Secretary of State for Environment, Food, and Rural Affairs in the very heart of government.

The next day, the UK became the first country to declare a state of climate and ecological emergency.

We have already seen a huge shift in public opinion. More and more people are joining this movement as they realize that the climate crisis – and the associated crises of capitalism and colonialism that caused it – will not be solved by gradual reform and rotten compromise. This is a crisis that requires radical system change on a scale never seen before.

Extinction Rebellion is a decentralized mass movement of concerned citizens. It is open to anyone who takes action in a non-violent way, actively mitigating for power and standing by the action that we have taken. We work to transform our society into one that is compassionate, inclusive, sustainable, equitable and connected; where creativity is prioritized and where the diversity of our gifts is recognized, celebrated and encouraged to flourish.

We believe that government has failed to understand the severity of this crisis. We believe that we must now take radical action to reduce the very worst effects of climate breakdown and, in doing so, reform and extend our broken democracy. We therefore have three key demands:

1/ the government must tell the truth by declaring a climate and ecological emergency, working with other institutions to communicate the urgency for change

2/ the government must act now to halt biodiversity loss and reduce greenhouse-gas emissions to net zero by 2025

3/ The government must create and be led by the decisions of a Citizens' Assembly on climate and ecological justice

These demands have been adapted by different international groups in accordance with our decentralized system of governance. The tactics we use in the United Kingdom or the United States are not always effective or safe in other countries, especially those under repressive regimes or dictatorships.

The majority world needs no lectures from us. They have been on the front line of this struggle for centuries and we do not presume to tell them how to stage a rebellion. We act in solidarity with them and their struggle, and we bow to their experience and their wisdom. After all, this is a movement led by indigenous communities and those in the majority world. It is led by women and children and people of colour. It is led by the people most affected by this crisis – the people who are being displaced now, the people who are dying now, the people who have been trying to warn us for years.

We acknowledge that Extinction Rebellion is just one articulation of a feeling that is being felt all across the world. We see ourselves as one branch of a much wider, stronger, wiser movement.

The future will not look like the present. We know that for certain. The rebellion will not happen under any one banner or any one slogan. The future is going to be humble. Because, if this is going to work, then we will all have to work together. After all, we are facing an unprecedented global emergency and our governments have completely failed to protect us. To survive, it's going to take everything we've got. And everyone we know.

The challenge we now face is extremely daunting. Because the problem, unfortunately, is not just the climate. The problem is ecology. The problem is the environment. The problem is biodiversity. The problem is capitalism. The problem is colonialism. The problem is power. The problem is inequality. The problem is greed, and corruption, and money, and this tired, broken system.

The problem is our complete and utter failure to imagine any meaningful alternative.

Perhaps this book will go some way to changing that. We need to rewild the world. That much is obvious. But first we need to rewild the imagination. We must all learn how to dream again, and we have to learn that together. To break down the old ways of thinking and to move beyond our current conception of what is and what is not possible.

This book is supposed to be a handbook. A book that you will keep by you, that will help you, inform you, empower you to act. A book that will compel you to join the rebellion in whatever way that means to you.

The first section of the book is about telling the truth; it will spell out the severity of the situation and describe, in painstaking detail, the effects of climate breakdown. It will tell you the facts and it will not hold back. It presents dispatches from the front lines of climate change and attempts to diagnose decades of denial. It considers the psychological damage of the climate crisis and the role of love, grief and courage in finding a way out of the wreckage. The second section is about action; it will give you practical instructions for what to do now and how to react. It will give you the tools to be an activist and gesture towards what happens next.

We know that, for some, this book will be very hard to read. You may end up feeling sad, or empty, or guilty, or angry, or frightened, or numb. But, ultimately, this is a book about love. About the love we should have for humanity, and the love we should have for the planet.

About the love we are currently lacking. And the love that we desperately need.

With love and peace. Rebel for life.

PART ONE

TELL
THE
TRUTH

IN TIMES OF UNIVERSAL DECEIT, TELLING THE TRUTH IS A REVOLUTIONARY ACT

— George Orwell

1/ DIE, SURVIVE OR THRIVE?

FARHANA YAMIN

At this point in human history we have three choices: to die, to survive or to thrive.

From the wildfires in the USA, coral die-back in the tropics and the deadly hurricanes battering small islands, the signs are crystal clear: climate devastation is already here. The world's poorest people and indigenous communities are on the front line. They are also bearing the brunt of the sixth mass extinction, which is under way due to conversion of their forests, wetlands and other wild landscapes into concrete cities, dam reservoirs and fields growing soya.

I joined Extinction Rebellion to fight against the climate and ecological emergency we are now facing – an emergency that threatens the very conditions of all life on Earth. I have been an environmental lawyer for thirty years, working to create new treaties, EU agreements and national laws aiming to prevent the situation we now find ourselves in. Sadly, I know this emergency cannot be averted by governments signing weak compacts and voluntary agreements with the biggest polluters on Earth. Nor by tweaking carbon markets that have been gutted of climate ambition by fossil-fuel lobbyists. We need to overhaul our political systems to limit access to government by big business. We need citizens' assemblies to allow ordinary people to decide the scale and pace of transition on the basis of independent scientific advice.

We need new laws to prevent ecocide – the destruction

of the insects, plants and ecosystems that humanity needs for food, pollination, clean water and healthy oceans. Every parliament, state legislature and local authority needs to declare a climate and ecological emergency, following the lead of ninety-one councils in the UK, including London and the UK Parliament, and to start allocating resources differently.

The reality is that politicians and powerful elites who benefit from 'business as usual' are not going to stop their destructive practices or loosen their grip on the financial and economic levers. They will keep asking for fossil-fuel subsidies. The official estimates of financial support to fossil fuels are between US$ 370 billion and 620 billion over the period 2010–2015, with the UK spending £10.5 billion a year, making the UK the biggest fossil fuel subsidiser in the EU. They demand we acknowledge their unfettered right to carry on financing new coal mines and fracking, and opening up gas and oil reserves, even in the last remaining pristine places on Earth. We must insist that our rulers acknowledge that human rights and ecosystems are protected under law and must now be put at the centre of our legal, political and economic systems.

Normal politics has failed us. It has brought the whole planet to the brink of ecological disaster. We cannot invoke and rely on the inadequate legal tools of the past thirty years that have allowed this crisis to happen. We need everyone to unite – from the left, the right, and every shade in between, and especially young people, many of whom are too disillusioned to vote or are excluded because they are only 16. We need everyone to undertake mass civil disobedience to create a new political reality the whole world over.

But we can't get there if we work in silos and factions. We need a 'movement of movements' to model the unity and urgency we need right now. The new movement of movements must be led by our youth and by those who have been resisting 'business as usual', especially communities of colour and those at the forefront of oppression. The new movement must be based on the reality that the legacy of colonialism, combined with current forms of capitalism based on never-ending extractive growth, is literally killing us. The reality is that four environmental defenders a week are being killed in the Global South. We in the Global North need to honour their work and join their struggles by also throwing our own bodies on the line.

We need a socially just transition for everyone on Earth, not just for workers trapped in the toxic industries that need to be phased out, but for everyone at the sharp end of austerity and ecological destruction. Climate and ecological destruction are at heart issues about social and intergenerational justice. We can't just fixate on the maths and science of climate change and leave people and fairness out of the equation.

These are the facts that justify the unification and intensification of our shared struggles and which will only succeed if we have a worldwide rebellion. Climate-change denialists cannot cover up the fact that the struggle for access to natural resources, especially fresh water and arable land, is intensifying, and that large parts of the planet are already becoming uninhabitable due to food and water scarcity.

Humans have transformed 51 per cent of Earth's land cover from forest and grassland to crops, cities and grazing lands, but in ways that undermine agricultural productivity, destroy biodiversity and encroach on indigenous lands.

Topsoil is now being lost ten to forty times faster than it can be replenished by nature, and 30 per cent of the world's arable land has become unproductive due to soil erosion since the mid-twentieth century.

The world's insect population has fallen by 60 per cent since the 1970s. Large parts of Europe look green but are 'biodiversity deserts' – the birds and bees are dying. Current extinction rates are at least tens, and possibly hundreds, of times greater than background rates, destroying entire eco-systems both on land and in the sea.

Climate change is warming up the atmosphere, oceans are acidifying and the cryosphere – the parts of the world covered in ice – is literally in meltdown. Abrupt, non-linear, irreversible changes are underway in the Arctic, Antarctica, Greenland and the world's glaciers, which are crucial to food, water and agricultural production.

The human consequences of these changes – economic instability, large-scale involuntary migration, conflict, famine and the collapse of economic and social systems – are plain to see and reported daily, but these stories are not linked in mainstream political and media coverage to the climate and ecological emergency that is already upon us.

Between 2006 and 2011, 60 per cent of Syria suffered the worst long-term drought and crop failures in the country's history. Two to three million people became poor and many more were internally displaced. The resulting social instability amplified the political factors that led to war in Syria, with now half its original population of 13 million having migrated or been internally displaced. Something similar is occurring in Yemen, where up to 10 million people face starvation, despite millions trying to move to safer, once fertile areas.

The received political wisdom that people in rich countries can sit tight and buy their way out of catastrophic environmental outcomes, or know that the welfare state will save them, is looking more and more fanciful as we remain in the grip of austerity politics. Anyone with an understanding of how the global food system works, especially how much of the world's food supply passes through less than a dozen 'choke point' ports, will know that our economies are deeply intertwined. Everyone will be affected, joining the millions who already are all over the world. Poor communities, especially people of colour, whether in the Global North or the Global South, who have always been on the front lines of environmental injustice, will likely also bear the brunt of the new catastrophes.

Are humans destined to become extinct as a species? Will we be slugging it out for what little remains by arming ourselves and building walls to keep out those less fortunate than ourselves? Can we really dismantle the toxic systems that have given rise to these gargantuan problems in the short window we now have?

No one knows what will happen, and no one can say for sure whether or not fundamental ecological tipping points have already been breached. The good news is that there are millions of people mobilizing to stop humanity falling off a cliff.

And they also have some sharp new ideas to create kinder, regenerative societies that can start the process of restoring nature and create communities of resistance and resilience to the impacts we cannot avoid. They want to do more than just avoid extinction or merely survive. They are building a movement built on solidarity and well-being so everyone, and every part of everyone, can flourish. Ending domination over nature goes hand in hand with tackling all forms of domination and hierarchy. The struggle for climate justice is also the struggle for racial, gender, sexual and economic equality.

We are not alone. Actions are now happening all over the world.

In Ghana, Extinction Rebellion activists recently held an event calling for action on the climate and ecological emergency in Africa. The event was staged in solidarity with those in the Global South and the entire world. 'The impact of the climate catastrophe is part of our daily life. This is why this event is important,' says the Ghanaian activist Mawuse Yao Agorkor.

These voices cannot be ignored any longer. This year, Extinction Rebellion has injected a new sense of energy and urgency into the climate movement. Thousands of people have joined, participating in non-violent actions by blocking bridges, blockading roads and shutting down government buildings.

While media headlines have focused on our work in the United Kingdom, Extinction Rebellion has started an International Solidarity Network to support existing resistance in the majority world, working closely with activists in West Papua, Bangladesh, Mongolia and the Caribbean.

Extinction Rebellion is also linking with and learning from other movements.

At the UN Conference on Climate Change, Extinction Rebellion supported the Alliance of Small Island States and the Climate Vulnerable Forum – together representing over eighty countries with 1 billion people. We helped pull together an international 'emergency coalition' to reject weak language that would have condemned them to extinction. While we in the Global North might only just be feeling the effects of climate change, the majority world has long since known the tragedy that the climate crisis brings.

Support is also being provided to the youth-led school strike movement started by Greta Thunberg, and to the newly emerging Birthstrike movement which is taking off in many countries to support people who are choosing not to bring children into this world unless, and until, conditions improve. In the US, the Sunrise Movement is building bi-partisan support for a ten-year mobilization and investment plan called the Green New Deal.

What all these movements have in common is a complete rejection of neoliberal economics and 'business as usual' politics. Yes, it is too late to prevent all the negative impacts of climate change. But this cannot destroy our capacity to nurture. It cannot destroy our capacity to love and our sense of justice.

We can and now must redesign human societies based on love, justice and planetary boundaries so that no person or society is left to face devastating consequences and we learn to restore nature together.

Faced with toxic systems that are destroying all life on Earth, affirmation of this vision and rebelling against whatever gets in its way becomes a sacred duty for all. We can and must succeed in catalysing a peaceful revolution to end the era of fossil fuels, nature extraction and capitalism.

Life on Earth depends on it.

2/ SCIENTISTS' WARNINGS HAVE BEEN IGNORED

PROFESSOR WILLIAM J. RIPPLE AND NICHOLAS R. HOUTMAN

'We stand now where two roads diverge. But unlike the road in Robert Frost's famous poem, they are not equally fair. The road we have long been traveling is deceptively easy, a smooth super-highway on which we progress with great speed, but at the end lies disaster. The other fork in the road – the one 'less traveled by' – offers our last, our only chance to reach a destination that assures the preservation of our earth.'

– Rachel Carson, *Silent Spring*

The job of scientists is to tell the truth, but the bearers of bad tidings don't always fare well. It matters little if they bring news from the past or warn of dangers in the future. Our efforts in the face of political and social pressures pursuing 'business as usual' have been belittled, ignored, and worse.

One warning stands out for its eloquence and impact – Rachel Carson's *Silent Spring*. Her masterful narrative about the disastrous overuse of DDT and other pesticides generated strong reactions when it was published in 1962: editorials in newspapers, invitations to testify before Congress and, not surprisingly, legal threats from the chemical industry.

While scientists largely applauded her work, some

powerful politicians and business leaders openly baited her on the basis of gender and loyalty to the nation. Although struggling with the cancer that would eventually kill her, Carson never wavered. She remained composed and confident in her message. Scholars credit *Silent Spring* with spurring the environmental movement in the United States and leading to policy changes, including the formation of the US Environmental Protection Agency in 1970.

Two scientists' warnings issued thirty years after *Silent Spring* focused on global warming. In 1990, the Union of Concerned Scientists (UCS), a non-profit organization in Cambridge, Massachusetts, published the *Appeal by American Scientists to Prevent Global Warming*, which the group called 'the most serious environmental threat of the 21st century'. Two years later, UCS produced the 'World Scientists' Warning to Humanity', which was signed by more than 1,700 scientists, including a majority of living Nobel laureates in the sciences. It describes clear evidence of global warming, species extinctions, freshwater declines, soil degradation and other problems.

The second World Scientists' Warning to Humanity, for which I was the lead author, was published in the journal *BioScience* on the twenty-fifth anniversary of the first. We reviewed trends in many of the areas identified in the first warning. In our top-ten list of planetary threats, our statement included charts of global data on the oceans, forests, freshwater resources, vertebrate species and atmospheric carbon and temperature. We concluded that, with one exception, conditions have worsened since 1992.

As retired NASA scientist James Hansen writes in *Storms of My Grandchildren*, scientists tend to be reticent in communicating the implications of their findings, a result perhaps of their adherence to the scientific method. 'Caution has its merits,' he wrote, 'but we may live to rue our reticence if it serves to lock in future disasters.'

We scientists have been frustrated and even in despair over the many years of inaction, but we will continue to speak out, telling the truth about what we all need to do to protect life on planet Earth.

A new network, the Alliance of World Scientists, is growing in the wake of this second warning. Our statement, co-authored by scientists on six continents, has since been co-signed by more than 21,000 scientists from 184 countries, making it one of the most cited science articles ever. Our warning is yet another effort to use science to advocate for human welfare and environmental sustainability – to share what research reveals about our likely future.

We are advocating for evidence-based solutions to the emerging planetary catastrophe, following in the footsteps of other warnings issued by scientists before us. In the decades after the Second World War, researchers conclusively linked tobacco smoking to lung cancer and called attention to the hazards of nuclear-weapons testing to the atmosphere. They identified industrial compounds as the culprits in destroying protective ozone in the stratosphere.

Today, those who issue cautionary statements have been targeted by opponents more concerned with profits and the status quo than with the health of people and the planet. The stakes for humanity have become global and the warnings have never been more dire.

3/ WE ARE NOT PREPARED TO DIE

MOHAMED NASHEED

President of the Maldives, 2008–12

Although we are vulnerable, we are not prepared to die.

The Maldivian people have no intention of becoming the first victims of the climate crisis. We are going to do everything in our power to keep our coral reefs intact and our heads above water.

We harbour no illusions about the dangers climate change poses. For the Maldives, climate change isn't an environmental issue. It is a national security threat. It is an existential emergency. The Maldives are disappearing. We will soon be under water.

The recent IPCC (Intergovernmental Panel on Climate Change) report is crystal clear: emissions must be reduced by 45 per cent in twelve years to stabilize global warming at 1.5 degrees Celsius. That is a daunting task.

And climate change is already upon us: weather patterns are changing; coral reefs are dying; erosion and water contamination are getting worse. But that doesn't mean we are going to give up. We plan to survive in a warming world, any way we can.

Low-lying atoll nations, such as the Maldives, are on the climate front lines. Coral nations will be hit first by climate change, and hit hardest.

The coral reefs that protect our islands and provide us with food are threatened with extinction by ever-higher ocean temperatures that bleach and kill coral. Then there is the longer-term threat of submersion by the rising seas – the highest point in the Maldives is just six feet above water.

In our quest for climate adaptation, we don't want to concrete over coral reefs to make a sea wall. We need to work with nature, not against it. That's why we need soft but smart adaptation strategies. We will build seawalls that encourage coral reef growth. We will grow mangroves to protect ourselves from stronger storms. We will use the latest science and cultivation techniques to grow corals that can survive hotter and more acidic seas, because some warming is inevitable, even if we commit to zero emissions today.

Of course, we cannot adapt for ever, as the world gets hotter and hotter and hotter. All nations, big or small, rich or poor, are in this together, whether they like it or not. It is just madness for us to allow global carbon dioxide levels to go beyond 450ppm (parts per million) and temperature increases to shoot past 1.5 degrees Celsius. That can still be prevented. If we come together on the basis of the emergency facing us, the world can do it. If we don't, we will all have hell to pay.

The Maldives is not prepared to allow that to happen. This year, as part of our commitment to the UNFCCC (United Nations Framework Convention on Climate Change) process to stop climate change, the Maldives will strengthen its nationally determined contribution under the Paris Agreement, committing aggressively to adopt solar power and renewable energy.

I am heartened that other nations, including those as vulnerable as we are, are planning similar things. The Marshall Islands, through its leadership of the Climate Vulnerable Forum, has become the first country to submit new targets under the Paris Agreement. Many other vulnerable, and industrialized, nations have also committed to drive fossil fuels out of their economies. This is welcome and hopeful news.

I have spent much of my life trying to bring freedom and democracy to the Maldives. When I was elected president in 2008, I started fighting against climate change. The fact I was ousted in a *coup d'état* in 2012 made it clear to me: the fight for democracy and against climate change are one and the same struggle.

How many autocrats are dependent on oil revenues to prop up their violent regimes? How many of those who damage established democracies are also climate deniers?

Neither can we ignore the recent violence on the streets of Paris, where working men and women protested against a modest increase in fuel tax. Here's the thing about climate change: we cannot frame it as a war between working people and saving the planet. If we do that, we will stir up the forces that led to the wave of populism that has engulfed the West, and some of the East, too.

Let us not forget what we owe to decent, working people such as coalminers. The tremendous wealth the world enjoys today, the technological progress, the huge increase in living standards, is due to the work of these people. We should not blame coalminers, or loggers, or oil-rig workers, for causing the climate crisis. Instead, we should thank them, for helping to fuel human civilization. Coalminers are not the problem. Coal is the problem. And we confuse the two at our peril.

When it comes to acting on climate change and shifting the global economy off fossil fuels, I propose a test: the working people who stand to lose most from the end of the fossil-fuel age should be the first to gain from the new, clean economy. And we should apply that test before each and every intervention we make.

When we can march through the streets, hand in hand with the miners and the oil-rig workers in a protest for climate action, then we will have unlocked the politics for transformative change.

We Maldivians are a nation of survivors, and we will do everything we can to ensure the survival of our country. But we can only survive as a nation if we also survive as a planet.

4/ THE HEAT IS MELTING THE MOUNTAINS

KAMLA JOSHI AND BHUVAN CHAND JOSHI

Interviewed by Adam Hinton for a film commissioned by Project Pressure

We live in a village in the Almora district, in the Indian Himalayas. We have two daughters and one son. Our village has around twenty families.

Our daily routine is farming. Get up in the morning, take a bath, worship God. Go to the cowsheds, take out milk. Go back inside the house and make tea. At eight o'clock, after breakfast, the children go to school. Whoever has to go to the farms will go there; the one walking the cattle goes to the forest. At two in the afternoon, everybody gets together for an hour and then, after eating, whoever has to do sowing and digging will go to the farms and the ones at kitchen work will go home, gathering firewood. We do this together. We come home, the children come back. Then just cooking and eating the food in the evening. This is the daily life here.

The village is very old. People who are eighty say that it is older than them, which means that we have been living here since the time of their great-grandfathers. Around 150 years. They all farmed here.

It was easier to farm then. Time would be in our favour, for sowing seeds. It would rain on time in every season. In our lifetimes, this is now less than half as likely as before. It has also been raining more during the monsoon than it used to. The weather we want doesn't happen.

The biggest daily problem is the wild animals. It is a big threat. They terrorize the farms. There never was much danger of wild animals: they weren't here at all. Their numbers have increased. They destroy the fields, cause so much damage to farming. The cutting of trees and building of roads is a reason for this. Cutting the trees makes all the flowers and jungles fall.

It used to be that, when monsoons came and clouds thundered a lot, there were so many trees that the rain would be soaked up by the trees themselves. Now the monsoons create destruction everywhere. Some places get flooded, there are landslides, houses break, there is mud everywhere in some farms. If there is a crop, it would spoil. This is the biggest issue.

We are not able to prepare much for this. We don't have a good way of protecting ourselves from the weather. There is no way to save our crops, nor do we have the resources. We don't even get any help to come up with a solution. Because if the crop goes bad, then all our time is spent producing the second crop, preparing it and seeding, tilling soil. This is why we are unable to prepare much.

It's nature's calling, so there's no prevention. There are mountains here. Whatever it is, it's all on nature.

Rain breaks the roads, electricity and water. For many days communication lines are broken. Houses float away. The houses are weak; if it rains too much, if the land is weak, then everything breaks down.

The government also reaches out very late to help but is not able to do so. We call them for so many days for help as everything is broken down, but they are not able to reach us in time. This is why there are a lot of deaths on mountains because of rains. You see this in the newspapers. Every time it rains, emergency is declared, teams are created. Resources for disaster management don't reach here.

It was never this hot. Bodies of water would not dry up before. There is no medium to irrigate and crops get destroyed. This heat is melting the mountains. It has started becoming hot there, too.

This is how we are supposed to keep going on.

Why is the weather changing? There must be some problem with the environment that has caused the weather to change.

Trees are cut down in huge numbers, which leads to increase in heat on the glaciers, which leads to their melting. The entire world heating is the main reason. If not the world, then at least on the mountains. Because the increase in temperature is one of the main causes of the melting of the glaciers. Scientists say this is so.

The spreading of diseases in the mountains is also due to the melting of the glaciers. The spread of malaria nowadays, the breeding of mosquitoes, is because of the melting of the glacier. All the germs and insects are born during the rain only. Farming also takes place during the rain, otherwise there is no water here. Most of the work also takes place during the rainy season. All the crops are grown and kept for winters. It won't rain in the winter.

Snow used to fall every year. This year it was less than before. Previous years, it snowed a lot, but not this time. I don't know what is happening.

Before all this, it was possible to grow enough at home for six months, eight months, nine months, so for two to three months men would do labour work here and there and buy the necessary groceries, whatever. Now, if after a lot of difficulty you produce some harvest, then there is food for three months only. For the remaining nine months, you have to buy it from outside, like from a cheap store or a private general store. To manage a family with these expenses becomes difficult. We get out of one problem and get into another; farming doesn't flourish, then houses need repairing, then there is no time for farming. You are troubled from both sides.

If farming does not happen, then there will be no produce; automatically, it will be expensive. But there is no employment here.

Migration is the only option.

If you keep getting your daily bread, then no one would want to leave their home. But people have nothing, so migration has become the biggest issue.

In the last ten to fifteen years, more than 50 per cent of the people in our village have migrated. People like me that are still left, people who don't have money or anything, who can't leave, will end up starving.

Either we migrate, or we drag ourselves to death. If we go hungry, we will suffer more illness.

What can we do?

We just get stressed. That's it. Stress is all we have. There is a shortage of everything. But we can't migrate. Where would we go?

We have to stay here. I don't know what the future holds. I only wish for my children that they study and learn something, become something or plan something in some good place. Study and find a good job. That's it. I only have this wish. For my children.

— Bhuvan Joshi, Binalal Pandey, Inder Singh Jarout, Kamla Joshi

5/ FIGHTING THE WRONG WAR

JS RAFAELI WITH NEIL WOODS

A plane skims low over the Colombian rainforest. It's a small, one-propeller Cessna. The pilot keeps a nervous eye out for anti-aircraft fire from guerrillas in the jungle canopy below.

The plane is spraying herbicide to eradicate coca leaves, grown by desperately poor farmers to supply the insatiable demand of the Western cocaine market. The guerrillas fire at the plane to protect the crop that finances their brutal para-military campaigns.

The herbicide the plane is spraying is called Glyphosate. It is manufactured by the American company DynCorp and sprayed as part of a US-financed Colombian military campaign to suppress coca production. DynCorp, along with the US and Colombian governments, insist that Glyphosate is non-toxic for humans. Yet in 2013 Colombia agreed to pay Ecuador $15 million in compensation for severe health problems the spraying was causing to families along the border. Despite multiple lawsuits, DynCorp itself has never paid a penny in reparations. Glyphosate isn't just killing people. The Colombian Amazon is perhaps the second most biodiverse habitat on the planet. We are wiping out species we do not yet know exist. This is the ecocidal face of the global war on drugs. This is just one of many intersections between climate change and the war on drugs.

*

We'll begin with the police. Since 2010, British activist circles, and society at large, have been rocked by the so-called 'Spy Cops' scandal. Undercover officers infiltrated environmental campaign groups, acted as agents provocateurs, formed intimate relationships with female activists, fathered children with them – and then abandoned their new families once their deployment was over. This was a systemic failure of policing ethics on a monstrous scale.

The police learned these dirty tricks while fighting the war on drugs. Undercover policing is itself a tactic of the drug war. Its widespread use in the UK emerged in the 1980s, specifically as a response to the growth of drug gangs. Neil, one of the authors of this piece, spent fourteen years as an undercover drugs cop. He was actively encouraged to form sexual relationships with vulnerable female targets. His point-blank refusal to do so led to strained relations with his police handlers.

Just as Neil's experiences turned him against the drug war, many of the most innovative and progressive alternatives to strict prohibition are now, surprisingly, coming from within the police themselves. Filling the gaps left by failed policy, it is increasingly police and crime commissioners who are instituting opiate substitution therapies, pill testing and diversion schemes. The international organization Law Enforcement Action Partnership is spearheading the realization that the war on drugs is, in fact, a betrayal of the police's original moral purpose – summed up in Robert Peel's founding principle: 'The police are the community, and the community are the police.'

There's something to be learned here. Extinction Rebellion has chosen a very particular stance vis-à-vis the police: to actively try to get arrested. The police will have to choose what stance to adopt in return. Despite appearances, law enforcement is not a monolithic ideological block. There are serious ethical voices there. A space may emerge in which law enforcement can be engaged in conversations around their own supposed founding principles.

The image of the American plane spraying toxic chemicals on to poor farmers below throws a harsh light on just how rich-world consumption habits can affect poorer 'producer countries'. It also highlights a perverse discourse around personal ethical choice and creating systemic change.

When confronted with rising crime in inner cities or the plight of countries like Colombia and Afghanistan, Western politicians frequently respond with a tired mantra about how 'middle-class cocaine users are fuelling an evil market and must examine their consciences'.

There's half a gram of truth in this. No one consuming illicit drugs should be under any illusion about the hideous conditions of their production.

But there is one overriding factor that makes those conditions inevitable – the structure of prohibition. It is the global war on drugs that prevents the regulation of the products themselves and the conditions of their production and distribution. Politicians weaponize conversations around personal consumption specifically to obscure the systemic ethical catastrophe of the global drug policies they support. The hypocrisy is grotesque.

We see the same dynamic in climate discussions – the outsourcing to the individual to solve systemic problems, the familiar clichés that this will all be solved by individuals recycling, buying reusable coffee cups and possibly downloading an app.

Personal ethical choice is undeniably important, for both the drugs trade and the environment. But as thinkers and campaigners, we cannot let these discussions be weaponized to derail the push for systemic change. Those who created these problems will try and make them yours to solve – don't let them.

Of all the lazy journalistic clichés used to speak about the environment, one of the most common is that our economy is 'addicted to fossil fuels'. The word 'addiction' here means something. It emerges from very specific discourses in medicine and criminal justice. In its modern context, it emerges from the war on drugs.

When dealing with people suffering from 'addiction', there are, historically, two dominant treatment philosophies. The first is abstinence. Here, the overriding goal is to make the user stop using, no matter their circumstances. Drugs are considered a moral evil so, by definition, the object must be to get people 'clean'.

The other guiding principle is known as harm reduction (HR). In this tradition, drugs themselves are seen as morally neutral. The idea is to reduce the potential harm to those who use them, and to society at large.

Harm reductionists support initiatives like needle exchanges to prevent the spread of HIV, or opiate substitution therapy: supplying users with methadone or heroin, allowing them to regulate their lives without the constant stresses of needing to score. To harm reductionists, the danger is not the drugs themselves. The harm comes with users being forced into poverty, criminality, sickness, prostitution and death.

If our economy is indeed 'addicted' to fossil fuels, perhaps, rather than punishing it, we might begin to conceptualize a harm-reduction approach to help it make the changes it urgently needs.

Harm reduction is practised all over the world, but it was

born in the UK, on Merseyside, as a response to the heroin crisis of the 1980s. Dr Russell Newcombe, who first theorized HR as a coherent system, outlined four key principles.

The first is pragmatism. As a drug worker, you must realize that not everyone is able to achieve abstinence. It is your responsibility to meet people where they are, not where you wish them to be.

The second is to remain non-judgemental. Drug users do behave badly sometimes – but they are still human, still somebody's son or daughter. Stigmatizing attitudes, from drug services in particular, are a major factor in why many users choose not to engage in treatment at all.

Next come requirements to remain user-friendly and relevant. Harm-reduction workers are expected to make their services as easy to access as possible, and to give users what they actually need – not what the drug worker or society at large think they should have.

How can we apply these principles to limiting climate change and engaging in activism around it?

If our economy is indeed addicted, we will need to meet it where it is, not where we wish it to be. Abstinence will not be achieved overnight – if, indeed, ever. But surely bold steps can be taken to provide alternatives and to reduce the harms this addiction causes, helping the user attain the stability they need to change.

Likewise, as thinkers and campaigners, we will encounter those who disagree with us. This can be frustrating, exhausting, infuriating. Yet, a harm-reduction approach requires us to remain non-judgemental. Stigmatizing attitudes cause people to disengage.

Searching for solutions that are accessible and relevant is emphatically not an argument merely for undue compromise or 'softly softly' approaches. If a drug user is in acute crisis – or in the midst of an overdose – then emergency interventions must be made. A harm-reduction approach to climate change would explicitly not be about letting the world heat up while taking small actions to mitigate the worst effects. It would be about taking bold, proactive and immediate steps to ensure that appropriate alternative structures are available as we urgently move to eliminate our carbon dependence – whether that is through massive subsidies to renewable energy or other creative measures.

Harm reduction is about looking truthfully at where we are as a starting point, helping others to confront the situation honestly – and managing the necessary transformations without undue trauma. Planes will probably continue to fly over the Amazon for some time – but not spraying poisons on the rainforest below.

In the face of the enormity of climate change, all other policy begins to feel like displacement activity. One day we will simply realize we have been fighting the wrong war.

6/ THERE'S FEAR NOW

FIREFIGHTER, CALIFORNIA

I work for the California fire service as a wildland firefighter. We deal mainly with vegetation fires, not so much structures or buildings. I've been on a Type 3 wildland fire engine doing rapid response and 'hose lays', working out in the middle of nowhere with a hotshot crew (a team of twenty highly trained firefighters), or hand crew, creating firebreaks to slow down the progress of fires. We'd be dropped off by helicopter, hike out into the woods and stay out there for days, sometimes weeks, removing anything that could fuel the fire in that area. We'd scrape the ground down to dirt and rock, mineral soil, and cut a line around it. Sometimes we'd create a large contingency area; sometimes we'd be right on the edge of the fire itself.

Climate change is a big subject when you're out in the field. A lot of the older guys who've been in the industry for forty years say the recent level of destruction is unprecedented.

The fire size, intensity and severity are increasing. Urbanization is sprawling further into drought-stricken wildlands. The trees are already dealing with bark beetles and pathogens and can't cope. They die, dry out. They're ready to burn.

You often hear people say, 'Well, we've been fine numerous times before, it's no different.' But now it is different. The preventative focus on vegetation management around homes, creating proper clearance, has not been as effective as previously thought . . . Embers jump into homes. There's a domino effect: all the houses go up.

There's a lot of finger-pointing over all this, but it's kind of pointless. This problem now affects whole towns. In Santa Rosa – an urban area – the Kmart burned down. It's getting closer.

I'm not emotional about burning structures: you get used to that. What's painful is the way fire affects people: when they're actively evacuating and scared and are told to leave everything behind because the fire is imminent. I see dead animals, horses, family dogs that were left behind.

If you're a millionaire in Malibu, you can rebuild. But communities like Paradise [a town swallowed by fire in December 2018] are mostly older, retired, working-class folks. They can't afford to bounce back.

To put this into perspective, the 2017 Thomas Fire in Ventura, California, burned two thousand homes. It was the most destructive in our history. In the last two years it has been surpassed four times. Nobody talked about it, but now they are. There's fear now.

Due to the political nature of the conversation around climate change in the US and a fear of raising potential employment issues, the author asked to remain anonymous.

THERE'S NO WAY TO PREPARE FOR THIS; IT'S INCOMPREHENSIBLE. IT FEELS LIKE THE APOCALYPSE; BLACK, BURNED

AND DECIMATED . . . A DEVASTATING EMPTINESS . . . LIKE A HYDROGEN BOMB . . . THE END OF THE WORLD

— Ted Silverberg, California resident, fire survivor

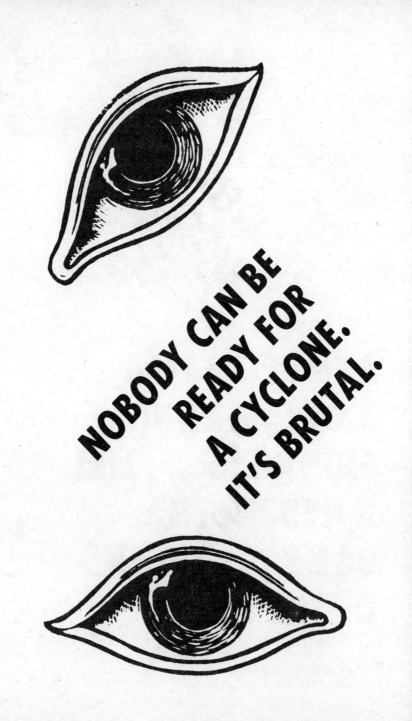

NOBODY CAN BE READY FOR A CYCLONE. IT'S BRUTAL.

WIND SPEEDS OF 180 MPH; THE EFFECTS ARE DEVASTATING, LIKE SOMETHING EXPLODED AND DESTROYED EVERYTHING

— Tsvangirayi Mukwazhi, photographer, reporting from Mozambique

7/ INDIGENOUS PEOPLES AND THE FIGHT FOR SURVIVAL

HINDOU OUMAROU IBRAHIM

When my mother was born, in the early 1960s, in the nomadic indigenous communities, Lake Chad – which gave its name to my country – was covering 25,000 square kilometres between Chad, Cameroon, Niger and Nigeria. This is the equivalent of the size of a small country in Africa, like Rwanda. Today, this lake, which is the heart and soul of the Sahel region, covers only about 2,500 square kilometres. In fifty years, we have lost 90 per cent of this resource so essential for the life of one of the poorest regions of the world.

Across the globe, grandmothers and grandfathers of indigenous peoples can tell us similar stories. They are living witnesses of climate change. The grandmother of my sister Jannie, from the Sámi people of the Arctic, can describe precisely the disappearance of ice and snow. The grandfather of my brother Cerda, from the Quechua people of Amazonia, can tell how the rainforest was once infinite, and how it's disappearing because of deforestation. The ancestors of my brother Samson of the Pacific see the coral bleach and the sea level rise gradually. Some, like my brother Gallego and others, even lose their lives defending their land to protect their communities.

Everywhere on the planet, more than 370 million indigenous people who live according to the rhythm of nature, in symbiosis with their environment, tell the same stories. These stories tell a tale about the end of the world.

Climate change is often something abstract for the inhabitants of rich countries, for all those who live in cities and who have forgotten what they owe to nature. But not for us. For us, it's a reality. A reality that comes from elsewhere. We are witnesses of the consequences without being able to act on the causes. Yet we do not want to resign ourselves to seeing nature die before our eyes. Because nature is our life. Because nature is our identity.

In the community where I was born families keep pace with the seasons and water to feed their huge herds of thousands of cattle, which help protect the fragile ecosystems of the Sahel through pastoral transhumance. Men read the wind, the clouds, birds' flight and insect behaviour to know when the rain will arrive, in order to guide their animals to pastures. Women find, in each season, plants that will cure diseases. Nature is our pantry, our supermarket, our pharmacy.

But in the last one or two decades things have been changing. Some livestock diseases, unknown until recently, have appeared. Water is scarce; pasture, too. Heatwaves up to 50 degrees Celsius kill humans and animals alike. Since the beginning of the century, my country has experienced a warming above 1.5 degrees Celsius, which is what the planet should know by 2050. In some parts of the world, for example the Arctic, indigenous peoples face a warming higher than 2 degrees Celsius. And by the time Europe perceives the first signs of climate change, we will already be living in a climate-changing era.

This change brings poverty; it also brings conflicts. The least water source becomes the object of all tensions, between fishermen, farmers and breeders. Men start to fight for water or fertile lands and weapons are never far away. Chad is far from Europe, where we saw, in the summer of 2018, helicopters arriving to supply water to herds hit by droughts. Here, if water does not come from the clouds, it will not come from anywhere else.

Water is not the only problem. Because with it, plants and animals essential for ecosystems and biological diversity disappear too. Women see their traditional medicines evaporate, preventing grandmothers from passing on to their daughters and granddaughters the names of plants that can cure diseases.

With this knowledge disappearing, it is a part of the memory of humanity that is threatened with extinction. All knowledge that is lost, every ancestor that goes out, is a book that burns, or a library that disappears.

Indigenous peoples are the guardians of ancestral knowledge that draws from the environment the solutions of everyday life. These solutions are priceless. It is a treasure for all those who have to face climate disorders, because it helps them to cope with the worst of its consequences, such as droughts, floods, hurricanes. In the Pacific Islands, indigenous peoples transmit from generation to generation the varieties of edible plants that can feed an entire people after a typhoon and so allow them to survive when all crops are destroyed. In the Sahel, the elders know which source continues to flow at the worst time of droughts. In tropical forests, indigenous peoples know which plants best protect against some epidemic diseases such as malaria or dengue fever.

Indigenous peoples are also the ones who best protect nature, because it is their work tool. In the heart of tropical forests, it is in the areas populated by hunter-gatherer communities that we can find the most biodiverse areas. In the Sahel, annual transhumance contributes to the natural fertilization of soils, thus developing a great green wall that prevents desertification. In coastal areas, from the Central American Kuna to the Pacific Maori, traditional fishing methods preserve corals, mangroves and other unique ecosystems that are the most effective barrier to rising sea levels.

Indigenous peoples do not want to be the silent victims of climate change. They are ready to share their traditional knowledge, and to (re)teach humankind how to live in harmony with nature. But for this, it is imperative that all major polluters in the world respect their commitment, made in Paris in 2015, to do everything to limit global warming to below 1.5 degrees Celsius.

Because if, in the next ten years, the international community does not resolutely engage with the transition toward a net-zero-emissions world, then I will likely be part of the last generation of indigenous peoples.

If tropical forests continue to be replaced by crops, if we are excluded from our lands, if the oceans are filled with plastics, if the chemical pollution that goes with industrial agriculture does not cease quickly, then our peoples will disappear, and with them millennia of knowledge of listening to nature.

8/ SURVIVAL OF THE RICHEST

DOUGLAS RUSHKOFF

Last year, I got invited to a super-deluxe private resort to deliver a keynote speech to what I assumed would be a hundred or so investment bankers. It was by far the largest fee I had ever been offered for a talk – about half my annual professor's salary – all to deliver some insight on the subject of 'the future of technology'.

I've never liked talking about the future. The Q&A sessions always end up more like parlour games where I'm asked to opine on the latest technology buzzwords as if they were ticker symbols for potential investments: blockchain, 3D printing, CRISPR. The audiences are rarely interested in learning about these technologies or their potential impacts beyond the binary choice of whether or not to invest in them. But money talks, so I took the gig.

After I arrived, I was ushered into what I thought was the Green Room. But instead of being wired with a microphone or taken to a stage, I just sat there at a plain round table as my audience was brought to me: five super-wealthy guys – yes, all men – from the upper echelon of the hedge-fund world. After a bit of small talk, I realized they had no interest in the information I had prepared about the future of technology. They had come with questions of their own.

They started out innocuously enough. Ethereum or bitcoin? Is quantum computing a real thing? Slowly but surely, however, they edged into their real topics of concern.

Which region will be less impacted by the coming climate crisis: New Zealand or Alaska? Is Google really building Ray Kurzweil a home for his brain, and will his consciousness live through the transition, or will it die and be reborn as a whole new one? Finally, the CEO of a brokerage house explained that he had nearly completed building his own underground bunker system and asked, 'How do I maintain authority over my security force after the event?'

The *event*. That was their euphemism for the environmental collapse, social unrest, nuclear explosion, unstoppable virus or Mr Robot hack that takes everything down.

This single question occupied us for the rest of the hour. They knew armed guards would be required to protect their compounds from the angry mobs. But how would they pay the guards once money was worthless? What would stop the guards from choosing their own leader? The billionaires considered using special combination locks on the food supply that only they knew. Or making guards wear disciplinary collars of some kind in return for their survival. Or maybe building robots to serve as guards and workers – if that technology could be developed in time.

That's when it hit me: at least as far as these gentlemen were concerned, this was a talk about the future of technology. Taking their cue from Elon Musk colonizing Mars, Peter Thiel reversing the ageing process, or Sam Altman and Ray Kurzweil uploading their minds into supercomputers, they were preparing for a digital future that had a whole lot less to do with making the world a better place than it did with transcending the human condition altogether and insulating themselves from the very real and present danger of climate change, rising sea levels, mass migrations, global pandemics, nativist panic and resource depletion. For them, the future of technology is really about just one thing: escape.

There's nothing wrong with madly optimistic appraisals of how technology might benefit human society. But the current drive for a post-human utopia is something else. It's less a vision for the wholesale migration of humanity to a new a state of being than a quest to transcend all that is human: the body, interdependence, compassion, vulnerability and complexity. As technology philosophers have been pointing out for years now, the transhumanist vision too easily reduces all of reality to data, concluding that 'humans are nothing but information-processing objects'.

It's a reduction of human evolution to a video game that someone wins by finding the escape hatch and then letting a few of his BFFs come along for the ride. Will it be Musk, Bezos, Thiel . . . Zuckerberg? These billionaires are the presumptive winners of the digital economy – the same survival-of-the-fittest business landscape that's fuelling most of this speculation to begin with.

Of course, it wasn't always this way. There was a brief moment, in the early 1990s, when the digital future felt open-ended and up for our invention. Technology was becoming a playground for the counterculture, who saw in it the opportunity to create a more inclusive, distributed and pro-human future. But established business interests only saw new potentials for the same old extraction, and too many technologists were seduced by unicorn IPOs. Digital futures became understood more like stock or cotton futures – something to predict and make bets on. So nearly every speech, article, study, documentary or white paper was seen as relevant only insofar as it pointed to a ticker symbol. The future became less a thing we create through our present-day choices or hopes for humankind than a predestined scenario we bet on with our venture capital but arrive at passively.

This freed everyone from the moral implications of their

activities. Technology development became less a story of collective flourishing than one of personal survival. Worse, as I learned, to call attention to any of this was to unintentionally cast oneself as an enemy of the market or an anti-technology curmudgeon.

So instead of considering the practical ethics of impoverishing and exploiting the many in the name of the few, most academics, journalists and science-fiction writers instead considered much more abstract and fanciful conundrums: Is it fair for a stock trader to use smart drugs? Should children get implants for foreign languages? Do we want autonomous vehicles to prioritize the lives of pedestrians over those of its passengers? Should the first Mars colonies be run as democracies? Does changing my DNA undermine my identity? Should robots have rights?

Asking these sorts of questions, while philosophically entertaining, is a poor substitute for wrestling with the real moral quandaries associated with unbridled technological development in the name of corporate capitalism. Digital platforms have turned an already exploitative and extractive marketplace (think Walmart) into an even more dehumanizing successor (think Amazon). Most of us became aware of these downsides in the form of automated jobs, the gig economy and the demise of local retail.

But the more devastating impacts of pedal-to-the-metal digital capitalism fall on the environment and the global poor. The manufacture of some of our computers and smartphones still uses networks of slave labour. These practices are so deeply entrenched that a company called Fairphone, founded from the ground up to make and market ethical phones, learned it was impossible. (The company's founder now sadly refers to their products as 'fairer' phones.)

Meanwhile, the mining of rare earth metals and the

disposal of our highly digital technologies destroys human habitats, replacing them with toxic waste dumps which are then picked over by peasant children and their families, who sell useable materials back to the manufacturers.

This 'out of sight, out of mind' externalization of poverty and poison doesn't go away just because we've covered our eyes with VR goggles and immersed ourselves in an alternative reality. If anything, the longer we ignore the social, economic and environmental repercussions, the more of a problem they become. This in turn motivates even more withdrawal, more isolationism and apocalyptic fantasy – and more desperately concocted technologies and business plans. The cycle feeds itself.

The more committed we are to this view of the world, the more we come to see human beings as the problem and technology as the solution. The very essence of what it means to be human is treated less as a feature than as a bug. No matter their embedded biases, technologies are declared neutral. Any bad behaviours they induce in us are just a reflection of our own corrupted core. It's as if some innate human savagery is to blame for our troubles. Just as the inefficiency of a local taxi market can be 'solved' with an app that bankrupts human drivers, the vexing inconsistencies of the human psyche can be corrected with a digital or genetic upgrade.

Ultimately, according to the technosolutionist orthodoxy, the human future climaxes by uploading our consciousness to a computer or, perhaps better, accepting that technology itself is our evolutionary successor. Like members of a gnostic cult, we long to enter the next transcendent phase of our development, shedding our bodies and leaving them behind, along with our sins and troubles.

Our movies and television shows play out these fantasies for us. Zombie shows depict a post-apocalypse where people are no better than the undead – and seem to know it. Worse, these shows invite viewers to imagine the future as a zero-sum battle between the remaining humans, where one group's survival is dependent on another one's demise. Even *Westworld* – based on a science-fiction novel where robots run amok – ended its second season with the ultimate reveal: human beings are simpler and more predictable than the artificial intelligences we create. The robots learn that each of us can be reduced to just a few lines of code and that we're incapable of making any wilful choices. Heck, even the robots in that show want to escape the confines of their bodies and spend the rest of their lives in a computer simulation.

The mental gymnastics required for such a profound role reversal between humans and machines all depend on the underlying assumption that humans suck. Let's either change them or get away from them, for ever.

Thus, we get tech billionaires launching electric cars into space – as if this symbolizes something more than one billionaire's capacity for corporate promotion. And if a few people do reach escape velocity and somehow survive in a bubble on Mars – despite our inability to maintain such a bubble even here on Earth in either of two multibillion-dollar Biosphere trials – the result will be less a continuation of the human diaspora than a lifeboat for the elite.

When the hedge-funders asked me the best way to maintain authority over their security forces after the 'event', I suggested that their best bet would be to treat those people really well, right now. They should be engaging with their security staffs as if they were members of their own family. And the more they can expand this ethos of inclusivity to the rest of their business practices, supply-chain management, sustainability efforts and wealth distribution, the less chance there will be of an 'event' in the first place. All this technological wizardry could be applied towards less romantic but entirely more collective interests right now.

They were amused by my optimism, but they didn't really buy it. They were not interested in how to avoid a calamity; they're convinced we are too far gone. For all their wealth and power, they don't believe they can affect the future. They are simply accepting the darkest of all scenarios and then bringing whatever money and technology they can employ to insulate themselves – especially if they can't get a seat on the rocket to Mars.

Luckily, those of us without the funding to consider disowning our own humanity have much better options available to us. We don't have to use technology in such antisocial, atomizing ways. We can become the individual consumers and profiles that our devices and platforms want us to be, or we can remember that the truly evolved human doesn't go it alone.

Being human is not about individual survival or escape. It's a team sport. Whatever future humans have, it will be together.

9/ CLIMATE SORROW

SUSIE ORBACH

It is a curious paradox. The more we are connected at a national and global scale, the less we seem to be able to take on the calamities that are brought to our screens. Flooding in the UK, the Arctic ice melts, the tsunami in Indonesia, the poisoning of water in Canada.

Five years ago, a flood in lower Manhattan knocked out the electricity, devastated New Jersey beaches. The activities of New York City residents ceased. Elevators didn't work. The food shops had no lighting or fridges. Hospitals were on back-up generators. My daughter sent photos of lower Manhattan as a river. With the water receding, the city returned to normal and, with it, for many, the awareness of what should have been the wake-up call receded, too.

How can we explain this curiosity, the fact of climate change being in our face and yet our capacity for denial? We know there are nefarious political players involved in disputing the evidence – evidence they know is incontrovertible for it has often come from their own reports, such as those from Exxon Mobil in the 1970s. Such players change the language to soften what we hear and dump a surfeit of words and advertisements arguing that climate change is not proven. They aim to destabilize the certainties we know. They are always there to provide 'balance': either in the form of outright denial (and in the guise of how they are cleaning things up), or, more dishonest still, they contest the idea that fracking, deforestation and pipe-lines are polluting.

As these voices generate ever more distortions, we inadvertently accommodate them in some way. There are mechanisms inside of ourselves that allow us to cut off from what we know even as we separate our rubbish, take our shopping bags to market, watch our screens aghast, and endeavour to limit our footprint.

I've been puzzling this much as I puzzle over other forms of denial. We participate in activities that are often against our self-interest. We are seduced into thinking that uncom-fortable things will go away or that 'science' will solve the problems. But it's not accurate and the urgency upon us means we need to engage with our own denial.

How can we do this? What is required of us psychologi-cally to engage with rather than cut off from this knowledge? How can we envision what is happening when it isn't right in front of us? It's difficult to imagine one's own death. How much more impossible to imagine that human activities might mean extinction?

Among North Americans polled at the end of last year by the Yale Program on Climate Change, 73 per cent said they believed it and 69 per cent said they were worried. This is an eight-point increase since March 2018, so consciousness is changing. But the politics aren't. They are going backwards and the capacity to hold on to what we know and want seems slippery. We know and we don't know.

To come into knowing is to come into sorrow. A sorrow that arrives as a thud, deadening and fearful.

Sorrow is hard to bear. With sorrow comes grief and loss. Not easy feelings. Nor is guilt, nor fury, nor despair.

Climate sorrow, if I can call it that, opens up into wretched states of mind and heart. We can find it unbearable. Without even meaning to repress or split off our feelings, we do so. I am doing so now as I write. Staying with such feelings can be bruising and can make us feel helpless and despairing. It is hard, very hard, to stay with, and yet there is value in this if we can create contexts for doing so.

The feminist movement taught us that speaking with one another allows truths to enter in and be held together. In creating spaces to talk, we transformed our isolation and, although we have not focused our energy on issues of extinction, we need to do so now. We need to take that practice, to create spaces in which we can share how difficult this hurt is and how to deal with our despair and rage.

Facing feelings is not a substitute for political action, nor is it a distraction from action. Feelings are an important feature of political activity. Acknowledging our feelings – to ourselves, to one another – makes us more robust. We need to mourn *and* organize. It should not be one or the other.

We know from conventional political struggles that the less we understand emotionally, the more our potential victories will lead us to missteps and a weakening of our legitimate concerns. When setbacks and external manipulations occur, which they inevitably will, there can be a pull to manage difficult feelings by collapsing into sectarianism. We can find ourselves projecting our frustrations, fury and disappointments at the slowness of change on to those we mostly agree with rather than those responsible for endangering social justice and planetary conservation. We need to work in broad coalitions where differences can be tolerated rather than fracture effective political interventions. This doesn't mean weakening our positions or not having sound leadership. Leadership is critical, and we need a leadership that finds ways to encompass the range of progressive activities while speaking to people's emotional upset as well as their hopes.

Climate emergency is too important for us to go into the splits that can haunt progressive struggles. We have superb political analysis. What's missing is how to hold the feelings we fall into denying. If we look at how moved and concerned children are when they hear about endangered bears, we see a tap root for political action. That tap root should be part of the toolkit for activists. We need to accept our own feelings of grief and fear and we need to provoke conversations that touch the hearts of others. In doing so we will build a movement that can handle the horrors we are facing, without the secondary issue of internal denial. We will be more, not less, robust. More, not less, effective. More, not less, compelling.

10/ THE CLIMATE EMERGENCY AND THE END OF DIVERSITY

MATTHEW TODD

Over the last thirty years, scientific reports of wildfires and floods have by and large enabled us to make a bargain with ourselves: 'I'll settle for some bad weather over giving up burgers and flying.'

During that time, too, the West has undergone staggering social change. In 1987, when I was a struggling teenager, according to the British Social Attitudes survey 75 per cent of the British public said they believed homosexuality was 'always' or 'mostly' wrong. Fanned by a hostile media, violence against gay and lesbian people was common, people were legally fired from their jobs because of their sexuality and others were barred from funerals of long-term partners by homophobic parents. Most of the media was homophobic. Some was explicitly racist.

Racism, sexism, homophobia, transphobia and other issues are still pernicious problems (trans people are the current media punchbag), but society has changed massively. We have, thankfully, become significantly more civilized: the concept of 'equality' is mainstream in the West; diversity is taken up by major companies; footballers wear 'rainbow laces' to show support for gay people; schools – with some exceptions – teach kids that everyone should be respected; the MeToo movement has shaken society. The accepted view is that our society is progressing and will continue to do so.

Yet there seems to be very little concern about climate change and the ecological crises we face. When I post about, say, a bed-and-breakfast owner who has turned away gay people, there will, understandably, be thousands of tweets about it. But post about the breakdown of the natural world, and few engage with anything like the same passion.

However, none of this is guaranteed. In 2014, the first gay woman to come out in British public life, Maureen Duffy, on receiving an award from gay magazine *Attitude*, said, 'We can never accept that all is well. We've come a long way . . . but don't take your eye off the ball.'

Unfortunately, many of us who are concerned with social justice and identity politics, including the wider left-wing movement (as well as, of course, the right), have made what is looking every day more like a fatal mistake. We have not given any thought to how the express train of ecological breakdown will smash through this delicate diversity we have spent so much time building brick by brick.

We have forgotten that all of these important issues – in fact, every issue – resides within the most important issue bar none: 'the planet'. With a broken planet, we will have no gay rights, no feminism, no respect for trans people, no attempt at fairness and justice for people of colour. What we will have is a fight to survive and a lot of violence.

It's only recently that voices such as that of British broadcaster Sir David Attenborough have talked of the collapse of civilizations and societies, or what food insecurity will mean for us, and for generations to come. In February 2019, Extinction Rebellion's Roger Hallam put it bluntly: 'War, mass mental breakdown, mass torture, mass rape.'

In all this, our relatively new societal values will be threatened. Those who have had to fight hardest for their rights – gay men and lesbians, trans people, people of colour, women, and those who have traditionally taken up their fight – will inevitably once again become prominent targets.

After all, there is still extreme hatred of 'difference' out there. For instance, on the day same-sex marriage was brought into law in England and Wales, comments in an article in one mainstream British newspaper suggested that 'we need a civil war to stop it'. The rape threats that any women of profile receive online, and the racism that is so common, speak to something that has been lying dormant in the murky depths of our society but is now stirring again.

Brutality is only kept at bay by the rule of law and by there being a critical number of educated people, in work, healthy and with enough money and food to keep them invested in society. When people cannot feed their families, then the façade of law and order evaporates. When Sir David Attenborough talks of the collapse of civilizations, this is what it means: violence that most of us in the privileged West cannot even comprehend.

There is a terrible precedent. Berlin in the 1930s had a flourishing queer community. A man called Magnus Hirschfeld campaigned for rights at his Institute of Sexual Science and conducted the first gender-reassignment surgeries. Then came economic crisis and the Nazi rise to power: Hirschfeld's institute was ransacked, his books and research burned, gay men were put on lists, arrested, imprisoned and some sent to concentration camps. Millions of Jews, Roma and others were murdered.

These are the most appalling of times, made all the worse by the fact that most people seem to have no idea of how bad things really are.

Already, the flames of the far right are being fanned. The climate crisis is petrol on this fire. As a gay person, I am terrified. This keeps me awake at night. It's one of the reasons why I was at the launch event of Extinction Rebellion, was on the bridges in November and in the streets in April and why I will continue to encourage all people, especially my LGBTQ brothers and sisters and all who are concerned about social justice, to join us. If we don't, then, for gay people, the bad days of the past will soon be with us again – and that may be the least of our worries.

11/ DOOM AND BLOOM: ADAPTING TO COLLAPSE

PROFESSOR JEM BENDELL

Our climate is changing rapidly, destroying lives and threatening our future. We must act now to reduce harm and save what we can. In doing so we can rediscover what truly matters. That may seem less of a rallying cry than 'This is our last chance to prevent disaster.' But I believe it is more truthful and will be more lasting. It will also invite less disillusionment over time and help each of us to prepare. After all, when harvests collapse, we won't be eating our placards. We will be relying on the love we have for each other and the ways we have prepared.

Scientists and activists have been shouting for the past fifteen years about the imminent disaster we are creating. The latest message is 'We've only got twelve years' to prevent a disastrous 1.5 degrees Celsius of warming, but I'm not swayed any more. My reading of the latest data is that climate change has gone too far, too fast, with too much momentum, so that any talk of prevention is actually a form of denial of what is really happening. It is a difficult conclusion to arrive at. And a difficult one to live with. We have too little resilience in our agricultural, economic and political systems to be able to cope. It is time to prepare, both emotionally and practically, for a disaster.

I am a social scientist, not a climatologist. So who am I to spread panic and fear when the world's top scientists say we have twelve years? Like many readers, I had assumed the authority on climate was the IPCC – the Intergovernmental Panel on Climate Change – but it turns out they've been consistently underestimating the changes. In 2007 they said an ice-free Arctic was a possibility by 2100. That sounds far enough away to calm the nerves. But real-time measurements are documenting such rapid loss of ice that some of the world's top climate scientists are saying it could be ice free in the next few years.

Sea-level rise is a good indicator of the rate of change, because it is affected by many factors. In 2007, satellite data showed a sea-level rise of 3.3 millimetres per year. Yet that year the IPCC offered 1.94 millimetres a year as the lowest mark of its estimate for sea-level rise. Yes, you're right: that's lower than what was already happening. It's like standing up to your knees in flood water in your living room, listening to the forecaster on the radio saying she is not sure if the river will burst its banks. It turned out that when scientists could not agree on how much the melting polar ice sheets would be adding to sea-level rise, they left out the data altogether. That's so poor, it's almost funny.

Once I realized that the IPCC couldn't be taken as climate gospel, I looked more closely at some key issues. The Arctic looms large. It acts as the planet's refrigerator, by reflecting sunlight back into space and by absorbing energy when the ice melts from solid to liquid. Once the Arctic ice has gone and the dark ocean starts absorbing sunlight, the additional global warming blows the global two-degree warming target out of the window.

The implications even of small changes are immense for our agriculture, water and ecosystems. Even just one warmer summer in the northern hemisphere in 2018 reduced yields of wheat and staples like potatoes by about a quarter in the UK.

Unlike other years, the unusual weather was seen across the northern hemisphere, with declines in rain-fed agriculture reported across Europe. Globally, we only have grain reserves for about four months, so a few consecutive summers like 2018 and the predicted return of El Niño droughts in Asia could cause food shortages on a global scale.

Our civilization would struggle to hold itself together under such conditions. I hear many voices fending off despair with hopeful stories about technology, political revolution or mass spiritual awakenings. But I cannot pin hopes on those things. We should be preparing for a social collapse. By that I mean an uneven ending of our normal modes of sustenance, security, pleasure, identity, meaning and hope. It is very difficult to predict when a collapse will occur, especially given the complexity of our agricultural and economic systems. My guess is that, within ten years from now, a social collapse of some form will have occurred in the majority of countries around the world.

Having worked for over twenty-five years in environmental sustainability, I find it hard to accept that my career has added up to nothing; my sense of self is shaken because I had believed humanity would win in the end. We had been walking up a landslide. I find myself regretting all the times I settled for small changes when my heart was calling for large ones. I've grieved how I may not grow old. I grieve for those closest to me and the fear and pain they may feel as their food, energy and social systems break down. Most of all I now grieve for the young, and the more beautiful world they will never inherit.

This realization meant that I began feeling the impermanence of everything in a far more tangible and immediate way than before. My attention had always been fixed in the future but now arrived in the present, and I became aware as never before of other people and animals – of love, beauty, art and expression. I was reminded of what my friend with terminal cancer had said about his experience of gratitude and wonder, and of the peculiarly intense quality of our last meeting.

I am hardly the first to notice this phenomenon. The Russian author Dostoevsky described the delicious intensity of the last moments before his false execution. I believe we all need to go through such a process, individually and collectively. Putting all our hopes in a better future allows us to make compromises in the present, while letting go of a better future can allow us to drop false hopes and live in the present with more integrity. It might even make our activism more effective.

If we are to stop the rapid extinction of species and avert the possible extinction of our own, what might we do, as publicly engaged citizens?

If societal collapse or breakdown is now likely due to climate change, might we communicate that view as widely as possible without offering a set of answers and action agendas? There is a lot that people can gain from feeling lost and despairing before then piecing things back together for themselves, in their personal, professional and political lives. Without loving support of any kind, a sudden realization that collapse is now likely or inevitable in the not-so-distant future can trigger some ugly responses to difficult emotions.

My view is that normalizing discussions about how to prepare for and soften collapse will benefit society. Only collective preparations have a serious chance of working. Deep adaptation to climate change means asking ourselves and our leaders these four questions.

First, how do we keep what we really want to keep? As we seek resilience, our capacity to adapt to changing circumstances is crucial to survive with valued norms and behaviours.

A likely collapse in rain-fed agriculture means that governments need to prepare for how to ration some basic foodstuffs and enable irrigation systems for crops such as potatoes. It's unclear how our financial markets will respond to the realization of climate shocks: there is a risk that our systems of both credit and payments could seize up. Governments need to ensure we have electronic means of payment outside of the private banking system so trade can continue in the event of a financial collapse. Some responses for resilience will take a bit longer. We need to try to buy more time. Many geo-engineering ideas are highly dangerous and impractical. But one makes sense right now. We should be seeding and brightening the clouds above the Arctic immediately, as a global emergency response. We need to be reacting as we would if an Armageddon-sized meteor was hurtling towards Earth.

Second, what do we need to let go of in order not to make matters worse? People and communities will need to relinquish certain assets, behaviours and beliefs: withdrawing from coastlines, shutting down vulnerable industrial facilities, giving up expectations for certain types of consumption. There will be the psychological challenge of how to help people who experience dread, grief and confusion. Many of us may be deeply affected by the falling away of our assumption of progress or stability. How do we plan our lives now? That will pose huge communications challenges if we want to enable compassionate and collaborative responses from each other as much as possible. Helping people, with psychological support, to let go of some old attachments and aspirations will be important work.

Third, we need to explore the restoration of attitudes and approaches to life and organization that our hydrocarbon-fuelled civilization eroded. Examples include rewilding landscapes so they provide more ecological benefits and require

FRUGALITY
HUMILITY
EMPATHY

less management, changing diets back to match the seasons, rediscovering non-electronically powered forms of play and increasing community-level productivity and support.

Fourth, as we contemplate endings, our thoughts turn towards reconciliation: with our mistakes, with death and, some would add, with God. We can also seek to be part of reconciliations between peoples with different political persuasions, religions, nations, genders, classes and generations. Without this inner deep adaptation to climate collapse, we risk tearing societies apart.

Bold emissions cuts and carbon-drawdown measures are still necessary, to reduce as much as possible the mass extinction and human suffering of climate change, but we must also prepare for what is now inevitable. This Deep Adaptation agenda takes us beyond mainstream narratives and initiatives on adaptation to climate change, as we no longer assume that society as we know it can continue.

Faced with these scenarios, some people react by calling for whatever-it-takes to be done to stop such a collapse. That is, to attempt whatever draconian measures might cut emissions and achieve carbon drawdown in case it might stop the disaster. The problem is that such a perspective can quickly lead to calls for those with power to impose on people without it, for the powerful to satisfy themselves that what they are doing needs to be done no matter what the implication for people's lives and well-being. It is now clear that there will be tough decisions ahead. But rather than suggest that we can sacrifice our values for a chance to survive, instead we can make universal love our compass as we enter an entirely new physical and psychological terrain.

I cannot honestly hope for a better future, so instead I'm hoping for a better present. I'm earning less money and instead I'm eating better and feeling better. I'm not compromising my truth, because I have nothing to lose. I'm sleeping more, enjoying more and loving more. In this sense, my life is not doom and gloom. Instead, both doom and bloom are part of my everyday experience.

In facing our climate predicament, I have learned that there is no way to escape despair. But there seems to be a way through despair. It is to love.

12/ NEGOTIATING SURRENDER

DOUGALD HINE

The place looks like an Italian monastery, all cloistered gardens and red-tiled rooftops. On a bright spring day you can get caught off-guard: stepping out on to the open walkway that links one building to another, you find the air two seasons colder than the view from the windows seemed to promise. We are a long way north of the Alps, in the small lakeside town of Sigtuna, thirty miles outside Stockholm.

There are advantages to the location. A few weeks after I moved to Sweden we had a friend passing through, one of the fiercest activists I ever knew, whose work has run from hacking together networks for the Syrian resistance to fighting for transgender people's right to exist. As we sat together in a patch of sunshine on a chilly April morning she stretched her arms and sighed: 'This is the one place I come where it feels like I'm back from the front line.' No country is without its front lines, but two centuries of neutrality have given Sweden a sense of peace that is striking by comparison to most corners of the world, and nowhere more so than in Sigtuna, a town whose street plan hasn't changed in a thousand years. In an upper room of one of the wooden houses that line its main street, in 1942 the German theologian Dietrich Bonhoeffer held secret meetings with George Bell, the Bishop of Chichester, bringing news to the Allies from the German resistance of its plot to overthrow Hitler.

For a hundred years the cloistered buildings of the Sigtuna

Foundation have been a place of meetings between worlds, where artists and priests and scientists gather on neutral ground. The Climate Existence conference belongs to this tradition: a gathering where the facts of climate change are not kept at arm's length, where we grapple with the ways that we are changed by what we know. This latest meeting took place over three days and, though it was only the start of May and the leaves had not long been on the trees, the temperature was rising to match the architecture, the first taste of a relentless summer when forests would burn and wells run dry.

At the end of the first day Kevin Anderson, deputy director of the Tyndall Centre for Climate Change Research, took to the stage. Before he was a climate scientist, Kevin built oil rigs for a living. He has the bluntness of an engineer, together with the moral clarity of a man who hasn't flown in many years, rejecting the logic of many in his field who justify their carbon footprints on the grounds of the importance of their work. His message was stark: to have a chance of meeting the goal agreed by governments in Paris, to keep climate change within a limit of 2 degrees Celsius, impossible things need to happen. Things beyond the bounds of what even the most progressive elements in mainstream politics have been willing to contemplate, even on their best days, over the past thirty years. The less bad news, he went on, is that a lot of things have happened over the past decade that weren't meant to be possible. He listed the banking crash, the Arab Spring, the rise of Jeremy Corbyn and the election of Donald Trump. The point is not whether we would welcome these developments, whether they represent a move in the right direction, but what they tell us about the nature of the times in which we are living: these are times in which impossible things happen, things which all the sensible voices whose job it is to tell us how the world works were busy telling us couldn't happen until they did, and in this there lies a dark vein of hope.

As I listened, I thought of something Vanessa Andreotti – professor of race, inequalities and global change at the University of British Columbia – had told us earlier that day. They have a saying where she comes from in Brazil: 'When there's a flood coming and the water's at your ankles, you can't swim. When the water gets to your knees, you still can't swim. But when the water reaches your butt, it's time to start swimming.' When things get bad enough, types of action that were previously impossible become possible.

These two thoughts about impossibility set me wondering: whereabouts in our societies is the water high enough already to start swimming? Because when it comes to climate change, it can still seem like it's only lapping at our ankles. Even against the backdrop of the fires and the drought, the conversations I heard in supermarkets and at family gatherings last summer were mostly: 'Isn't the weather amazing?' and 'Don't the farmers moan a lot?' and 'The government ought to buy more of those firefighting planes so we don't have to keep borrowing them from Italy!'

If I had to guess where the waters are highest, I'd say it's places like loneliness, mental health among young people and technology addiction. And of course, compared to what climate change means right now in Kiribati or Mozambique, these are the very definition of 'first world problems': crises of meaning, rather than 'existential crises' in the literal sense of the ability to subsist.

But what I'm asking is, where are the sources from which 'impossible' change might come, the points where things are bad enough right now in the societies whose ways of living need to change most in the next few years if the worst of climate change is to be averted? And my hunch is that the answer might lie somewhere other than the obvious places on which action around climate and behaviour change tends to focus.

17 November 2018. I'm watching footage shot on camera phones. Against a backdrop of famous monuments, protesters bring traffic to a halt, interrupting the business-as-usual of Saturday in a capital city.

The images come from both sides of the Channel. There's something uncanny about the emergence on the same day of these two movements, Extinction Rebellion and the Gilets Jaunes, the similarity of tactics, even the aesthetic coincidence of the fluorescent colours, the yellow vests in Paris and the yellow, green and pink flags on London's bridges. At first glance they look like mirror opposites, two sides ranged against each other in the battle for the future.

There are lessons here. The carbon tax on diesel may have been the last straw for the French protesters, but there had been plenty of other straws. When market incentives are employed to tackle climate change, they tend to fall hardest on the people already getting the hard end of the deal in a market society, whether at a national or an international level. The academic William Davies calls neoliberalism 'the disenchantment of politics by economics' – an attempt to get the processes of measurement, competition and price to do the work of remaking society, without the qualitative judgements and collective decisions of politics. In this sense, what went up in smoke in Paris was the fantasy of green neoliberalism. This is where the third demand of Extinction Rebellion comes in: a serious response to the climate emergency will require a radical democratic process, a transformation of our way of living in which we participate as citizens, not just as consumers.

The shared tactic of the roadblock already points in this direction, insisting on the urgent need to slow down, to bring the rush of business-as-usual to a halt so that we can start to have a real conversation about how we are going to live, how we are going to change our lives, given what we know about the mess in which we find ourselves.

There's a clickbait ad that keeps appearing in my Facebook feed for an organization that wants to plant 8 billion trees. I haven't checked them out, but let's assume that the people behind it are for real, that they are as scared as you and me by what is happening with the climate and that their over-simple story about how we can fix things is offered in good faith. What gets me is their idea of hope. Because, after a vivid account of the future towards which we are headed, they offer this proposition: if only we plant more trees, then 'instead of the post-apocalyptic dystopia . . . everything continues as usual'.

In this morning's news, there's a report that the number of young people in the UK who say 'life is not worth living' has doubled since 2009. It now stands at almost one in five.

Think of the word 'sustainability'. Whatever it once meant, it ends up meaning the project of sustaining as much as possible of our current way of living, only with wind turbines and electric cars. Like the commentators caught off-guard by Brexit and Trump, the mainstream proponents of sustainability fail to grasp how limited is the appeal of 'everything continues as usual'.

When I think about what is at stake now, there's a phrase that keeps coming back: this is about negotiating the surrender of our whole way of living.

How much of the economic activity that you see around

you could just go, overnight, and no one would honestly miss it, *if* this could happen without anyone going hungry or homeless as a consequence? Of course, that's a huge *if*, especially when – as in the UK – rough sleeping and dependence on food banks have already been rising for years. Yet this is what a surrender looks like: it's about how much of the organized activity of a society can be decommissioned, not by 2050 or 2030, or even 2025, but as soon as possible. The fact that David Graeber found out when he wrote about 'bullshit jobs' – that much of our activity shown up as GDP is widely recognized as pointless, including by those carrying it out – is beneficial.

To negotiate a surrender, you need a credible threat – and this is where the movement that began in London in November 2018 might look again at its strange twin across the Channel. Clearly, Extinction Rebellion would not seek the edge of chaos which drew so much attention to the Gilets Jaunes. Yet if this is truly a rebellion, then – in its own non-violent way – it needs to carry the kind of threat to the existing order that forced Macron to back down. When the big NGOs start talking about 'the next wave of climate-change protests', we should be alarmed, because if that's all this is, it will achieve as little as the waves that went before.

Yet there is another side to negotiating a surrender. The militancy that brings the existing order into question needs to be matched by the quiet places of conversation, away from the front lines, where unlikely partners enter into dialogue. This is the story of peace negotiations everywhere. It is also the spirit of the campaigners in the Irish referendum on abortion, who were willing, as Fintan O'Toole writes, to 'talk to everybody and make assumptions about nobody'.

Finally, if this is a surrender, it's a strange one, for there are no victors here. We are not equally implicated, for sure – but we know that it is our way of living that must be surrendered, and not only the lifestyles of Macron and his friends in the Davos set. Recognizing this, we might set the insights of military strategists and peacemakers alongside the understandings of surrender to be found in spiritual traditions or in the treatment of addiction. Between them, they could tell us that to surrender is to give up, to be humbled, perhaps humiliated, but with a chance – not a guarantee – that we may live to tell the tale.

PART TWO

ACT NOW

POWER CONCEDES NOTHING WITHOUT A

DEMAND

IT NEVER DID AND IT NEVER WILL

— *Frederick Douglass*

13/ COURTING ARREST

JAY GRIFFITHS
Extinction Rebellion

I was locked-on at Oxford Circus in the brilliant sunshine of
Easter Saturday. Arrest imminent. Lying there immobilized,
my hand locked with a chain inside a metal pipe with layers
of concrete and roofing felt, I felt my vulnerability intensely. I
could feel the heat of sparks flying around my head and smell
burning metal as the police cut me out. As a rule, pain fright-
ens me. What's more, I am a writer, and it was my right hand
that was in the lock-on; the police were operating an angle
grinder barely an inch from my fingers. Nearby, Hannah,
seven months pregnant, was also locked-on. We would be
among the last to hold out, together with a solitary fourteen-
year-old and a man in a wheelchair.

Two nights previously, I was hand-fasted with – and
to – my partner at Parliament Square. Our hands were held
together with love and superglue, the moon shining over
Westminster Abbey. Courting in the middle of a rebellion.
David Attenborough's climate-change programme was being
screened by Extinction Rebellion, the ghostly scaffolding
sheeting at the Palace of Westminster a backdrop used as a
projection screen. An XR banner – 'Beyond Politics' – was
flying in the trees.

Not just courting but courting arrest. We were dis-
appointed that night. The police had been arresting some of
those locked-on or glued on to the tarmac, but then they

packed up and went home for the night. Before they left they offered us all solvent to release ourselves to have a nicer night.

'Why do you *want* to be arrested?' asked one of the police officers at Oxford Circus. This question is at the heart of Extinction Rebellion. I had never in my life been arrested, for all the usual reasons: a desire to be law-abiding; concern about acquiring a criminal record; nervousness about a fine or imprisonment; fear of being isolated in a locked cell. For XR, I wanted to be able to take non-violent direct action without any of those fears. When you seek arrest, calmly and willingly, the idea of it is no longer a deterrent. The sting is gone. So is the fear, because the way to stop being scared of something is to actively attempt it.

One of the most powerful ways to bring about change is when people are willing to be imprisoned for non-violent civil disobedience. In the tradition of the suffragettes (my grandmother, wonderfully, was a Pankhurst, though she never said if she was related to 'the' Pankhursts) and the civil rights movement, courting arrest means moving from 'bystander' to 'upstander'. Standing up for something infinitely bigger and more important than you.

This is the self-sacrificial idea of arrest at the core of Extinction Rebellion's strategy, and it gives you strength from within. Ancient values are overtly resurrected in this Easter rebellion in London: the values of chivalry and honour, faith in life and being in service to Our Lady, Notre Dame, Mother Earth, the mother on whom everything else depends. Everything. As both Notre Dames were burning.

Seeking arrest means putting yourself on the line: rebels with a cause, gently disarming the arm of the law by linking arms with it. The rebels are acting as first responders to the crisis of the natural world, but we are all part of the emergency services now.

I had been among the first rebels assigned to take Regent Street, quietly surging across the lanes of traffic at eleven o'clock on the first Monday morning of the rebellion. Up with the night watch, a police officer stationed there came over to congratulate us on the organization (a well-being tent, free hot meals, a compost toilet, as well as that iconic pink boat stencilled 'Tell the Truth'). He paused, then said, 'You're standing up for something that needed to be stood up for. We all needed someone to do that. You are doing it. I totally support you.' Over the course of the week, I lost count of the number of police officers who said they agreed with us, letting us know in different ways. One police officer driving a van of arrestees was quietly whistling, 'I'm sorry, my friend'; another shared her food with a rebel in custody.

It took the police about two hours to get me and my co-rebel out of the lock-on. In a lull in the proceedings, one of the officers picked out my partner in the crowd and passed on to him my special request that he sing Leonard Cohen's 'Hallelujah' – 'And I've seen your flag on the Marble Arch and love is not a victory march.' He cupped his hands and sang it to me, and it made people cry. Yes, including me. At that time, Extinction Rebellion flags still held all across central London, from Waterloo Bridge to Marble Arch. A warm and unbroken hallelujah.

They say that, in the cells, the worst moment is when the cell door clangs shut. It was. I felt like crying. The solitude. The silence. But the acoustics are great in a small tiled room, so I sang instead. Sang badly. Sorry, all.

'When you are in the cells,' an XR friend of mine told me, 'remember you are choosing to be there. It makes all the difference.' It did. In an age which has banalized the word 'choice', reduced it to consumerist pap, we take it back. There is a choice. To take sides. To be on the right side of history.

To choose life over extinction. There is a choice to put one-self on the line, because words (and this is a heavy heresy for a writer) are not enough.

My friend's case went to court a few weeks ago. The judge gave her the most lenient sentence possible. Her solicitor, referring to the Paris Agreement, mentioned the Paris Commune by mistake, and apologized, saying she was a novice when it came to climate change. The judge looked at the solicitor sternly and said, 'None of us can afford to be novices when it comes to climate change.'

In the cells, the police checked on us all through the night, offering us food and drinks, and asking what they could get us, in acts of dear hospitality. While I was in custody an altern-ative government was being created, a People's Assembly at Marble Arch, where, indeed, a kind of Paris Commune was established. Now, the stakes are infinitely higher than ever they were in Paris 1871 – important though they were.

Here, now, unfrightened and non-violent, we are irre-pressible. The People's Assembly has moral authority and a transparency that the shrouded Westminster will never know. 'Tell the Truth' was the pink boat's slogan, so I will. Yes, I have been mightily scared of arrest. Yes, I am a bit feeble when it comes to going without my creature comforts. Yes, I am still daunted by isolation. But the truth is, I'll be doing it again very soon.

As we left the police station at five o'clock as Easter Sunday dawned, the police wished us well. 'You lot have rejuvenated me this week,' said one. 'It's been the nicest week of my work-ing life,' said another. One reached out and shook my hand. 'God bless. And good luck.'

This is sweet rebellion.

14/ THE CIVIL RESISTANCE MODEL

ROGER HALLAM
Extinction Rebellion

Extinction Rebellion did not come out of nowhere.

Unlike many of the spontaneous social-media-fuelled rebellions and uprisings in recent years, Extinction Rebellion has been carefully planned. For several years a group of academics and activists have been working on two main questions: Why have we failed so miserably to stop climate change? And how the hell are we going to stop it?

To answer these questions, we went to the library. We studied decades of work looking at organizational systems, collaborative working styles, momentum-driven organizing and direct-action campaigning. This research, alongside the site research we have carried out ourselves, has been invaluable to the development of our ideas.

People have now joined us from many fields, bringing with them lifetimes of experience and wisdom. This is now a movement of scientists, academics, lawyers, diplomats, councillors, campaigners, teachers, doctors, nurses, artists, writers, actors, graphic designers, psychologists, and many, many more. They have all volunteered their time, largely for free, and this has amounted to an incredible level of valuable input which anyone could estimate at hundreds of thousands of pounds.

We have brought together a huge number of people, from all across the world: a diverse movement, united behind one shared vision. A vision of radical system change.

We have to be clear. Conventional campaigning does not work. Sending emails, giving money to NGOs, going on A-to-B marches. Many wonderful people have dedicated years of their lives to all this, but it's time to be honest. Conventional campaigning has failed to bring about the necessary change. Emissions have increased by 60 per cent since 1990 and they are still going up, increasing by 2.7 per cent in 2018 alone.

Looking at that thirty years of appalling failure, the reason is clear. The rich and powerful are making too much money from our present suicidal course. You cannot overcome such entrenched power by persuasion and information. You can only do it by disruption.

So let's talk about disruption.

There are two types of disruption: violent and non-violent.

Violence is a traditional method. It is brilliant at getting attention and creating chaos and disruption, but it is often disastrous when it comes to creating progressive change. Violence destroys democracy and the relationships with opponents which are vital to creating peaceful outcomes to social conflict. The social science is totally clear on this: violence does not optimize the chance of successful, progressive outcomes. In fact, it almost always leads to fascism and authoritarianism.

The alternative, then, is non-violence. This option was, of course, important in the twentieth century, used successfully by the civil rights movement in America and the Indian independence movement. From all the studies, the message is clear: if you practise non-violence, you are more likely to succeed.

We call this the 'civil resistance model'.

There are many variations, but this is the general outline:

1/ First, you need the numbers. Not millions, but not a few dozen people either. You need several thousand: ideally, 50,000.

2/ You have to go to the capital city. That is where the government is, that's where the elites hang out and it's also where the national and international media are usually based. The truth is, they don't mind you doing stuff in the provinces. They do mind when you set up camp on their lawn, because they are forced to sit up and pay attention.

3/ You have to break the law. This is the essence of the non-violent method because it creates the social tension and the public drama which are vital to create change. Everyone loves an underdog narrative. It's the great archetypal story in all cultures: against all the odds, the brave go into battle against evil. Breaking the rules gets you attention and shows the public and the elite that you are serious and unafraid. It creates the necessary material disruption and economic cost which forces the elites to sit up and take notice. Common actions are simple ones: sitting down in roads; painting government buildings.

4/ It has to stay non-violent. As soon as you allow violence into the mix, you destroy the diversity and community basis upon which all successful mass mobilizations are based. The young, the old and the vulnerable will leave the space. So people need to be trained to stay calm and groups need to be assigned the role of intervening when tempers flare up. This needs to be organized, and non-violent discipline is rule number one for all participants.

5/ It has to go on day after day. We all know A-to-B
   marches get us nowhere – and the truth is, neither
   does blocking a capital city for a day. It's in the
   news and then it's over. To create real economic
   cost for the bosses, you have to keep at it. The first
   day or two, no one is bothered. After a few days
   it become 'an issue'. After a week it's a 'national
   crisis'. This is because each day you block a city the
   economic costs go up exponentially – increasing
   each day.

6/ Last but not least, it has to be fun. If we can't
   dance at it, it isn't a real revolution. The artistic
   communities need to be on board: it's a festival.
   We are going to show the media that we're not
   sitting around waiting to die any longer. We're
   gonna have a party. Obviously.

The key lesson about all structural political change is this:
disruption works. Without disruption there is no economic
cost, and without economic cost the guys running this world
really don't care. That's why labour strikes are so effective
against companies and why closing down a capital city is so
effective against governments. You have to hit them where it
hurts: in their pockets. That's just the way it is.

The central dynamic here is the 'dilemma' action. When
you create a dilemma for the authorities you open a space
of opportunity which was not there previously. Within that
space you can get noticed, speak truth to power, negotiate,
and more.

The authorities now have a serious dilemma: let people
continue to party in the streets, or opt for repression.

The authorities cannot let you continue – but, if they go
for mass arrests or use violence, then millions of people will

see it. It will be international news. It only takes 1 per cent of those watching and following the disruption to go, 'You know what, this is terrible, I am going down to support these people,' and its checkmate for the authorities, game over. The more people they take off the streets, the more come on to them. The fear is gone and the party goes on.

This has happened over and over again in the past hundred years. The arrogance of the authorities leads them to overreact, and the people – approximately 1–3 per cent of the population is ideal – will rise up and bring down the regime. It's very quick: around one or two weeks on average. Bang: suddenly it's over. Unbelievable, but it happens that way. Of course, the civil resistance model doesn't work every time but, vitally, it enables you to roll the dice. Emailing and marches don't roll that dice. Partying in the streets does.

It's interesting and important to note that the people who are most effective are often the least attached to the effectiveness of their actions. Being detached from the outcome, and in love with the principles and the process, can help mitigate against burn-out. People make decisions about life like this based on traditional virtue ethics, and it sounds old-fashioned, but these include things like duty, honour, tradition and glory. We work best when we know we are doing the right thing, and to be liberated from a system that has compromised your morality your entire life changes everything.

Finally, there is a twist.

We all know everything needs to change, everywhere, and soon. Gradual reform in one country will not do the trick any more. The good news is that you just have to get on with it in one country. If you create enough of a stir, then it all kicks off in other places – what the academics call the 'demonstration' effect. Check out previous global rebellious episodes: 1848, 1918, 1968, 1989, 2012. It starts somewhere, the news spreads and millions of people come out on to the streets around the world.

The lesson then is you don't wait until everyone is ready, because you'll be waiting for ever. You just need to go out and do it.

Rebellions are created because some people have had enough. They are over it and don't care if they are successful or not. It's sublime madness. It's the only thing which will save us now.

We recently watched a film about the Freedom Riders. There is a scene in it that blew us away when we first watched it. Black and white students are preparing to go on coaches down to the segregated South and break the law – to sit together side by side on the bus. They know this could lead to injury or even death. The interviewer asks one of the freedom riders if he is scared. The guy looks at the camera and says with great force: 'I'm ready.'

So, yes, we're ready, too. And so are millions of good people in this country and around the world.

See you on the streets.

15/ MOVEMENT BUILDING

PROFESSOR DANNY BURNS AND CORDULA REIMANN

It should now be obvious to everyone that vested interests only change when they are forced to do so. In other words, civil disobedience is necessary to change the attitudes and behaviours of both individual consumers and governments.

Back in 1990, we helped organize the largest campaign of civil disobedience in modern British history. With others, we mobilized more than 10 million people in a collective act of mass civil disobedience against the Conservative government's Poll Tax. The much-hated tax set a flat rate for everyone, meaning multimillionaires paid exactly the same as cleaners. In an act of resistance, British citizens *en masse* refused to pay. In doing so, we brought down the Thatcher government. We brought about change then, and we can do it again now.

But how did all of this happen? To persuade people not to pay their taxes was not an easy thing to do. People faced fines, arrest and the threat of imprisonment. A tangible sense of community solidarity had to be built. This meant that we needed a strategy for group building that was based on face-to-face human connection rather than being mediated through technology.

In those days, social activism was not mobilized through computers. There are many things we can do now that we couldn't do then, but while information can perhaps inspire rebellion, it cannot build the mutual trust and solidarity which enables people to sustain that rebellion. If you haven't taken action before and you fear arrest, then first of all you need to feel you are among people who are organized and know what they are doing. The more successful the movement, the greater the backlash from the state and the more important such solidarity will be.

A tangible example of how solidarity was built at scale can be seen in the work of the Bristol-based Easton Anti-Poll Tax Union. We concluded that people would only refuse to pay if they felt others around them would also do so. Once it had grown sufficiently, the Union organized a representative for every street in Easton. That person took responsibility for knocking on every door in that street with a survey. One of the key questions ran along the lines of 'If 75 per cent of people in your street say that they won't pay the Poll Tax, will you also not pay?' We collated the responses and then went back to every household with the results, at the same time inviting people to group meetings. This sort of personal trust-building is crucial to mobilizing committed and sustainable action. People quickly realized that if they were taken to court, they would be followed by wave upon wave of others.

In the climate and ecological emergency we now face we need to take that group-building process beyond the neighbourhoods to the places where people work, worship and spend their leisure time.

The most important lesson here is this: while it is possible to build activism with a centralized 'strategy', it is not possible to build a mass movement that way. People act on what they believe in. If the strategy deviates from their perspective even a small amount, they tend to lose motivation and fall away from the movement. What we saw in the anti-Poll Tax campaign was a huge portfolio of actions, ranging from letter writing to mass demonstrations. People did what they were passionate about. They stayed involved because they believed in what they were doing, and they had an active role. People learned what worked because the 'strategy' was allowed to emerge; in this way, they gradually converged around what was successful.

What seemed to be contradictory activities reinforced each other. That is, the petition and letter writers helped to give the movement a sense of 'mainstream support', while the demonstrators provided a cutting edge of protest and signalled that people were prepared to act. Although many of these individuals could probably not stand to be in the same room as each other, they reinforced the effectiveness of each other's actions.

Building a movement is quite different from building an organization, and our task as activists and organizers should be more akin to planting trees than giving orders. Our responsibility is to ensure that the earth is fertile, that the seeds are good quality, that they have water, that they are sheltered enough to grow and that there are bees to pollinate them. If we do this, they can spread, and we will most likely not know exactly where or how.

The other important point to take away is this. Those who have vested interests in protecting the current system will adapt to protect it as fast as we adapt to expose and to challenge it. But we now have a more differentiated, practical and theoretical understanding of how change happens than our predecessors did. We must not replicate the mistakes of the past. Instead, we should build a large and flexible movement which has the ability to adapt rapidly to what will be a very fast-moving process.

The stakes are higher now than ever before. This time, we are literally fighting for our lives.

16/ BUILDING AN ACTION

TIANA JACOUT, ROBIN BOARDMAN AND LIAM GEARY BAULCH
Extinction Rebellion

In Extinction Rebellion, we use non-violent direct action to raise awareness of the climate crisis, drawing on a long and noble history of peaceful protest and direct democracy. Generally, our actions can be broken down into three types of action. Most of our protests will achieve all three at some level but are often organized to focus on one over another.

They are:

1/ **Disruption** – to create disruption through mass civil disobedience, towards achieving our demands.
2/ **Outreach** – to tell the public the truth and bring people together at the protest, or through media.
3/ **Visioning** – to demonstrate the future we want to see through beautiful creative collaborative action.

It is important to give yourself one month to organize most actions, unless responding to particular political situations. This is especially important for our regenerative culture. We are trying to create a sustainable model of activism, as this is an ongoing rebellion. Despite there being many opportunities to respond to situations, it is sometimes wiser to use our

own timescales. Being constantly pulled in different directions and overworking to respond to every new crisis will
lead to exhaustion and burn-out.

Think about what you are hoping to achieve. Shutting
down one area is fine, but targeting several will cause the
most disruption – therefore, you have to choose your areas
for maximum impact.

Tell everyone what you are doing – there is huge power
in telling the government where you will be and when. This
is an act of pure defiance. We are too many to stop. We are
reclaiming our power.

Choose an area that will be the 'safer zone'. This is for
families with small children and others unwilling or unable
to be arrested or on the 'front line', if the term is applicable:
make the police aware that more vulnerable people will be
present there. On Rebellion Day 1, Westminster Bridge was
half shut down anyway and was not 'critical' (i.e. no need
to clear and arrest) so it was chosen for visual impact, and
Blackfriars was named the 'kids' bridge'.

Think about different levels of civil disobedience and try
to design actions so that, whoever turns up, there are ways
they can be included and involved. The decision to risk arrest,
or prosecution, is a personal one, and is of course affected by
your social position. There are roles within our movement
for people who don't, for whatever reason, want to take these
risks.

It must be stressed that Extinction Rebellion actions are
focused on creating disruption through committing civil
disobedience. Peaceful lawbreaking and acts of rebellion
shift the narrative to one of emergency, which changes what
people are willing to risk to change a system which is risk-
ing all life.

Recognize important areas where access needs to be clear – hospitals, fire stations – and plan accordingly. We are here to disrupt, not to cause harm. The five central London bridges that Extinction Rebellion blocked in November 2018 were chosen as they are routinely shut down for the London Marathon, so the city has a pre-existing plan for its emergency services around these sites.

However many areas you choose to occupy, they should all be treated as discrete, individual actions. Get all your leads for each area to create working groups; for example, five individual areas require five well-being coordinators. They are in charge of working out what is needed collectively, gathering and training volunteers and splitting them up evenly between the areas.

Create a web of information flow. Actions, areas, stewards, wellbeing points, information tents, etc. If we are to truly decentralize, the hierarchy must become a web. Instead of a top, there is a middle. That middle is breath, heart and information. It empowers instead of commands. We flow organically together, like a flock of birds with no one leader, moved by our instincts of togetherness. We trust that we are all here for one shared purpose: system change.

There are so many things to consider when planning and organizing an action. Take advice and do not rush into anything too soon. Know your areas of ignorance and seek support from those more knowledgeable and experienced than you.

Your responsibility is to try to keep the rebellion peaceful, safe and sustained.

17/ FEEDING THE REBELLION

MOMO HAQUE
Food coordinator

During the London rebellion of April 2019 we fed thousands of protesters holding four key sites simultaneously, for ten days. Keeping rebels fed and watered is of course key to sustaining actions; it's a big logistical challenge. The role of food in the movement, though, goes deeper than that.

A positive, inclusive relationship with food allows us to create communities and a vision of what is possible. Whether it's a cake or a hot meal, it's important to feed everyone who comes into contact with the rebellion: not only the rebels themselves, but the police, homeless people, commuters on their lunch breaks.

I am incredibly proud that we fed the city and, in so doing, showed that the society we need to build is possible. Visitors to the sites felt this, too, and wanted to get involved. Each site started off with just four cooks. By the end, we had twenty to thirty new recruits who'd joined in to help in whatever way they could: chopping veg, sourcing ingredients, washing up. People joined us off the street, coming in their lunch breaks, returning at the end of their working day. These were regular passers-by, inspired by what they saw, smelled and tasted. They grew the community, helping us keep the kitchens going for up to twenty hours a day.

Here are a few rules of thumb we followed:

Encourage and accept all contributions of time and effort and of food and ingredients – even if something's not to your taste or diet, someone will be happy to have it.

Eating together has been proven to raise happiness levels and increase social bonding. Bring a folding table and Thermos flasks and have a picnic in the middle of the road. Better still, set up a marquee or a tent – a focal point where food can be shared. The faster the kitchen or food area is up and running, even just serving tea, coffee or water, the quicker a protest site feels solidified. Conversations are started, friendships are formed. A road can be immediately transformed into a family.

Cook vegan: besides being good for the planet, it ensures maximum inclusivity and top health-and-safety standards; plus, it's easier to source the ingredients. With fresh vegetables and a quick turnaround, there's nothing that will go bad, even without refrigeration. Try not to use any nuts; keep it gluten free. Be careful to flag up allergens, though sesame and soya can be difficult to avoid.

A simple dish that can be scaled up easily:
Rice and lentils – soaked together for around three hours.
Add some spices, maybe cumin, turmeric, chilli, ginger, garlic, onion, salt.
Add lots of water, cook for an hour or so; when it's soft and mixed together, it's ready to serve.

Presentation and a mix of textures are important to meal planning. Add some salad, greens, something crunchy that looks good on the plate, alongside smudgier pulses and carbs. Something sweet for dessert keeps energy high. We offered cake or fruit, ideally apples: food that can be entirely eaten, meaning less waste to compost. We had almost no wastage during the rebellion. Everything got used.

If you've put together an action for hundreds, or thousands, there's a lot of planning involved. Preparing food on site means extra focus on health and safety. Cooking knives must be securely locked away when not being used and gas canisters carefully monitored.

A relationship forms between the protesters and their kitchen. It's comforting for protesters to know there's always hot food and a cup of tea available. We found it crucial to seek assistance from other groups who share values, expertise, people and logistics. Hare Krishna's 'Food for All' team were amazing, taking pressure off our main kitchens at peak hours, preparing food off-site, keeping it warm in well-insulated portable ice-boxes to serve right at the heart of the action.

Don't forget to factor in washing-up. Encourage people to bring plates, cutlery. Create a washing-up station with sponges and dishcloths. Keep it civilized: change the water when it's murky!

18/ CULTURAL ROADBLOCKS

JAMES AND RUBY

Any object placed in the middle of a road is an obstacle. However, without people around to take responsibility for its presence it can easily be ignored and removed. A large concave frame of reclaimed wood in the middle of a bridge is just that . . . until it's activated by skateboarders, at which point it becomes a mini-ramp, a meeting point, a media focus, an authentic, creative vision of how much better things can be – a space for cars transformed into a people-first festival.

Interested pedestrians want to take a closer look at a grand piano amplified by cycle-powered generators, to get a selfie with mobile papier-mâché sculptures, or sit by a row of trees in pots to eat the free lunch served by a pop-up kitchen. In doing so they reinforce the community and the obstruction, blurring a line between engaged audience and active protester.

With time, people feel emotionally attached to these road-block presences. They are defining cultural features of newly emerging micro-communities who will go to extreme lengths to protect them. This was particularly true of roadblocks that acted as creative platforms or stages: an open-sided van on the bridge hosting concerts and performers, and, of course, the Big Pink Boat.

Through weeks of planning, our Extinction Rebellion boat managed to make a surprise appearance in the middle of Oxford Circus as the centrepiece of a 'Tell the Truth' stage.

This theme was chosen as the site is just a few minutes' walk from the BBC headquarters, where the climate and ecological emergency is not yet front-page news.

Our first option was an incredibly heavy wooden boat, 32 feet long. This fell through when we realized that specialist trucks and driving licences would be required to bring it from a distant coastline to central London. With only days to go, we all started bidding enthusiastically on Ebay.

One tiny thumbnail image stood out from the others: its rightness sang through the screen. There was just enough time on the auction, and it was only a few miles from central London and in 'sail-away condition'. We bid determinedly.

'So, what are your plans for her?' the seller enquired when we went to collect in person. The boat had gone for a song, far less than he'd hoped, and he was being very decent about it. We couldn't let on. Spontaneously channelling the spirit of a veteran skipper of few words, one of our team knelt down, opened up the boat's sail bag and gave a strong sniff, as if checking for damp. This was enough to convince our seller of our maritime expertise. 'We'll be taking her on quite an adventure,' we assured him, and promised to send photos.

At 21 feet this was no superyacht. But what made it stand out was the long, fixed keel, which meant the boat sat high up on its trailer. The hull and its vital message would be always visible above the crowd, appearing to float on a sea of beautiful rebels.

We decided to name the boat after Honduran activist Berta Cáceres. XR is an international movement, and, while we are able to protest in relative security in the UK, four environmental campaigners are killed every week in the majority world. Berta was murdered for her activism, and we wanted to memorialize her incredible sacrifice.

The original plan was to paint the boat in colourful 'dazzle' camouflage, but it was decided that this would obscure the lines of the ship, and we wanted it to be unmistakeable. So the natural conclusion was to paint it bright pink from bow to stern. Green or blue were dismissed as too boaty; yellow or orange too unnatural. Pink is a calm colour, without being soft. Red is angry, pink is fun. It was only after painting it that we realized proudly that, when viewed from above, there was now something vaginal in its appearance.

Our team rehearsed the boat's delivery to Oxford Circus over and over again, but it was still an anxious time. We were running late and could see the crowds of protesters waiting by the side of the road, primed to jump out to create the protest. If they got over-excited and moved too soon we would be blocked out with the other traffic.

With victory in sight, a last-second check revealed that the light cable from the tow vehicle was still attached as the driver began to make his exit; only quick action stopped a catastrophe. The wheel clamp went down and our mast went up to an unforgettable cheer.

We had the crowd's attention; now we had to make sure we kept everybody entertained. We had curated DJs and talks alongside events, including an international solidarity day with speakers from across the world. The sound system and lights ran entirely on renewable energy: batteries were charged by the road-blocking solar array at Marble Arch, relayed to and from Oxford Circus by a team of XR couriers with trailer bikes. It was reliably loud.

When the schedule went awry or the police had us surrounded, we had to improvise. Our main MC was actually in charge of production for the site – she had never picked up a microphone before. But she stepped into the role and became intrinsic to the cultural character of the boat. The

crowd held people's assemblies expertly facilitated from the deck to decide our course of action and resistance.

For the first few days the police mostly let us be, but affinity groups from Bristol and Wales had coordinated to 'lock-on' underneath the boat, establishing that it wasn't going to move – something that proved vital when the police really tried to move in. By literally glueing themselves to the boat, those brave people – the famous 'barnacles' – gained us several extra days of safe anchorage.

Creating a show and a traffic-stopping photo opportunity got us plenty of global media coverage and took our 'Tell the Truth' message beyond the site, but it couldn't really capture the emotion and community around this roadblock. Each protest site developed its own unique culture and identity; boat people were different from bridge people. We sent the boat-seller a copy of a media piece headlined, 'How to Get Arrested and Influence People' above a photo of the *Berta Cáceres* in all her pink glory, together with a message, 'First leg of the journey complete!' We haven't heard back.

19/ ARTS FACTORY

MILES GLYN AND CLARE FARRELL

We're not making one of anything, we're making ten thousand. Patches, flags for a full-scale rebellion.

We're making new things from old. We don't need new; we already have all we need. Share ideas, share skills, share paint. Share clothes. Destroy consumer products, transform them into your own messages. Nothing is for sale.

This factory needs to be a community, and we try our best to be non-hierarchical. We start by setting the goals, showing everyone how to do the task allotted to them, giving them tools and, most importantly, the autonomy to do their own thing.

Diversity of producers is vital. Everyone brings something to the look of the rebellion: some are good with colours; some are more playful. The messaging may be precise but we need variety so it doesn't look static or formulaic. Our style must be dynamic, pushing forward. We avoid aggressive colours; no obvious left-wing reds or Conservative blues, no alienation or division. We are rebelling for a cause that affects us all so we must be actively inclusive.

Our main focus is at both ends of the day – preparing the print workshop so everyone has a good time and the tools they need. Sometimes it all descends into chaos, rushing around with buckets of ink, and everyone's printing on everyone and everything.

We eat together. We encourage everyone to bring

ingredients and cook together. When it gets really hectic, people end up sharing plates, dipping in, sharing mouthfuls. It gets intimate.

The anarchistic traditions of working with your hands and getting them dirty doesn't mean the outputs should be trashy. We'll hand-paint banners and make sure they're as sharp as possible, thinking about how this might look in photographs, making the message clear; getting the font just right – but the odd drip looks really good. We need to embrace imperfection in a world where everything is too shiny.

WHY AN XR-ARTS FACTORY?

We call ourselves the 'Autonomous Anarchist Arts Factory'. Among ourselves, since it describes what we are doing together. We're industrious for the rebellion.

We present a different way of living.

People buy things to make themselves feel good, judging themselves on social media, hunting for likes – if we can show them community, sharing food, sharing real human feelings, then it's not difficult to offer a better alternative.

We want to look unified; not uniform, but hand-made. We work to save money. We aim to use ecologically sound materials, preferably recycled or reclaimed. Sourcing this material in volume is a huge job. The work we make contributes to telling the story of XR, images that spread the message.

We embrace imperfection, mistakes, over-inking and under-inking, the human touch. We have a DIY attitude. It is so joyful to work together on this, one of the most important projects of all time.

RESEARCH AND DEVELOPMENT PHASE

We spent much time and effort making prototypes, testing them, showing them to the group to get feedback, testing all the types of inks, taking advice from master makers on the best techniques.

We like hi-vis: it signifies hazard, and activists are doing roadblocks, so there's a safety consideration. We talked a lot about imagery, and there is more variety on the way. We mostly used yellow, since the legal observers were going to be in orange. Many woodblocks were made over the months, and with each batch of prototypes things were changed and improved.

We had made lots of patches before. Given the time pressure, we just went for it, then improved the design for subsequent batches.

SOURCING SAMPLES AND SUPPLIERS

Start early so you get great deals and to avoid any hitches in production. It takes time to find suppliers, receive samples and make tests. The hi-vis comes from textile recycling factories: it's cheap but takes some time to source and de-brand.

PRODUCTION
PRODUCTION
PRODUCTION

RECRUIT A MAGNIFICENT SEVEN

You need a consistent team. We like two to five people working together per day, from a stable of seven. This is a manageable size if the workers are dependable. Often people drop in for a few hours after work, artistic reinforcements: this is great, especially because by this time we tend to be whacked, so we get to slow down, tidy up or cook for the team.

TELL EVERYBODY EVERYTHING

Then let them organize themselves. This frees up the coordinator to plan the future, as well as empowering the team.

BLACKBOARD WITH GOALS AND PROGRESS

Write targets, what's complete and what's still to be done. It's satisfying and ensures we tell everybody everything.

ROTATE JOBS

So that everyone stays fresh and people can learn new skills.

MAKE THE FACTORY A MICROCOSM OF THE MOVEMENT

XR: At some point the Rising Up core principles had seeped into us. We listened to anti-consumerist music and lyrics by Wladimir M. – 'Planet E': a continual reminder of our aims and ideals. We enjoyed making the Future Anarchist Utopia a reality, even if only for a few weeks in the AAA-Factory.

CHANGE THE LANGUAGE WE USE

When talking about the work we make, it only takes a little effort to remind people how many dozens of people were involved in making one patch, from getting funding, to admin, to outreach, to printing and attending protests, and so on. We remind each other we are all connected and working together, with love and grief and rage.

REFLECTION, RESEARCH AND EDUCATION

In between busy periods we continue, at a slower pace, to research and develop ideas both new and old. We teach each other skills and discuss what we want the AAA-Factory to be and how it should evolve.

WORKSHOPS AND OUTREACH

We continue doing workshops printing on people's existing clothing so the symbol and messages will be seen everywhere. More blocks are being cut with normal clothing and the human body in mind. The words and imagery are always expanding.

We make and think about:
- Patches, overprinting and why DIY is needed
- Wearing a sash
- Being a message – embodying resistance
- Reclaiming workwear, hi-vis as camouflage
- Saying things without words – everyone loves emojis
- Linocuts and woodblocks, the heritage of DIY print in revolutions
- Taking a pre-apocalyptic approach to style and respecting what exists
- We will never need another charity T-shirt ever again
- DIY as a way of making people earn sh

it through action . . .

20/ ONE BY ONE: A MEDIA STRATEGY

RONAN MCNERN

Media and Messaging coordinator, Extinction Rebellion

I want to start by considering how an idea becomes a reality, then progresses to a household name. I've worked in PR for many years: during that time I've introduced concept ideas for a start-up to an uninterested media, ideas that have drawn a blank in terms of a media response. Then, we get that first breakthrough piece of coverage, always the most difficult to get. With the door ajar, more interest follows, a trickle at first, then a deluge.

All this is true of our strategy for communicating the climate and ecological emergency.

The story Extinction Rebellion has to tell is one of universal importance. Its audience is, ultimately, as big as it gets: namely, everyone. Yet XR's strategy – initially, at any rate – is not to reach everybody. When designing our media campaign we had to start somewhere, so we started with those we felt were most likely to hear the call.

Our media messaging is based on research by Erica Chenoweth and Maria J. Stephan, which demonstrates that to achieve social change the active and sustained participation of just 3.5 per cent of the population is needed. It's that 3.5 per cent that we want to engage.

The first action that Extinction Rebellion did in the UK in October 2018, which emphasized XR's unique approach and distinguished it from previous movements, was to occupy

Greenpeace's offices briefly with an act of beautiful creative resistance, bringing cake, flowers and a love letter. The point of this was to say, 'Greenpeace, we love you, but we need to talk. There is an emergency and you have a role to play in this.' It was also an action aimed at those disillusioned with normal 'environmental' activism, who were ready to hear our emergency messaging.

We offered exclusive access to this action to a number of embedded journalists. They included Real Media (a key independent UK video news site), the *Guardian* (unrivalled for its reporting on climate and ecological issues), plus the Canary (one of the leading alternative news sources in the UK), as we wanted to reach out to people who our gut instinct told us were ready to hear the call. The Press Association, while not embedded, was also a priority for us: their wire stories get picked up across the media.

This strategy planted the seed for what was to come. XR's campaign has never been about trying to get a one-off piece of coverage. Rather, it's about ensuring clarity of message, and about building solid and authentic relationships with journalists, one by one, in order that they gain a thorough understanding of the common danger we face.

A key rationale for this strategy was to ensure that we communicated why protesters were engaged in civil disobedience. This way, we hoped to avoid media coverage being solely about arrests. Instead, it would be more contextualized, more nuanced, stating why protesters were willing to risk their freedom for this cause.

As a result, by the time of XR's Declaration of Rebellion on Parliament Square on 31 October 2018, we had begun to generate interest from other media and the public. From this point on there would be two main strands to our work: working in an involved way with those ready to hear the call;

and communicating more widely with other media using a mailing list for press releases and statements.

Around this time, individual journalists began to contact us in a personal capacity, to say, in confidence, that they were one of us, a rebel. My answer to these people has been and remains an ask: that they continue to do what they can from their own position of influence; that there are many like-minded journalists out there. And that, together, we can change the world.

The next key moment came prior to Rebellion Day 1 on 17 November 2018, a protest that saw XR block five bridges along the Thames in central London. We invited a larger group of journalists to a press briefing, including representatives from the BBC, the *Independent*, the *New Statesman*, as well as a host of independent and freelance photographers and filmmakers. One journalist present said this was the first time he had ever seen an activist group stand up in front of journalists in advance of an action, outlining what we were about to do for a cause. It was also, the journalist added, the first time they had heard activists stating that they were seeking to get themselves arrested.

This briefing led to greater understanding about XR's aims and methods, especially from the BBC (though it may not have been immediately apparent in their coverage). At this point, too, we assigned members of our team to key media in order to build closer links with them. Our relationships with Sky, ITV, *The Times*, the *Sunday Times* and others began to deepen, in turn leading to opportunities that would go beyond the immediate news.

Then, in December 2018, XR targeted the BBC in a series of disruptive actions around the UK. This was done in order to highlight the potentially transformative role of the BBC in generating awareness of the climate emergency.

We knew that many in the BBC would like to be able to report more on climate change: it was clear that our pressure from below could empower them to advocate for change within the corporation. As a result of that day of action, many BBC folks began to realize that XR's noisy protests provided a useful focus for news coverage – coverage that might break through to a wider audience across the country.

As that first wave of rebellion calmed, towards the end of 2018, it was time to pause, to make room for the thousands of new activists that wanted to get involved – regionally, nationally and around the world. Among them were journalists, who deluged XR with offers of help. And so began the planning for International Rebellion, a broadening of the movement.

In April 2019 we've seen more than a thousand people in London put their bodies on the line for the climate emergency. This spectacular sustained protest has enabled us to create news headlines around the world, in turn opening up a space for telling the truth about the climate and ecological emergency unlike anything before.

With so many news stories breaking in London, exclusives have remained a key part of our strategy, our core target media understanding that we will offer them all individual opportunities to get their own stories. So far, our strategy has, by and large, worked. Our media team has grown, as has the number of journalists we work with. We've had positive coverage in the UK and in international media.

A few years ago, I created a WhatsApp group for my own family. It changed how we communicated with each other, enabling us to be more present in each other's lives. The immediacy of WhatsApp has also played a key role in how we in the Media and Messaging team communicate among ourselves. The addition of a WhatsApp group for journalists interested in Extinction Rebellion news changed the relationship we had with journalists. One by one this became a very powerful means of communicating with media during this International Rebellion, with news stories breaking within moments of being posted to the group.

Just as with my family, Extinction Rebellion has been able to be more present and human to journalists, stepping away from the formality of some of the traditional means of communication. We also used the Signal app for more secure conversations when needed, amongst a whole range of other apps and platforms.

Now, XR has broken through to mainstream awareness, at least in the UK. But this is only the start. The challenge before us is to take the Extinction Rebellion story to every facet of life: across international and national media and the myriad to vertical media sectors (consumer, lifestyle, health, education, and so on). We are like that start-up that has got its breakthrough piece of media. Our task now is to ensure it wasn't a one-off.

We know that many of those in the corporate and mainstream media want change too. We invite those interested in getting involved in our Media and Messaging teams around the world to get in contact. To journalists we say: there is a place for you in the transformative change that needs to happen. Every one of us has a role to play if we are to face the emergency and to bring meaningful change and new purpose to society.

21/ GOING TO JAIL

CATHY EASTBURN

I was arrested for supergluing myself to the top of a DLR train at Canary Wharf, charged under the Malicious Damage Act of 1861, and held in prison.

I didn't know until the day whether I'd do it or not, but I knew what I was letting myself in for: after all, it's about getting arrested, to force government action. That's the point.

I'm a musician, fifty-one years old, with two teenage daughters. I've been aware of the issues around climate change for over thirty years now and have become increasingly concerned as the decades have slipped by. I've signed petitions, given money to NGOs, taken part in all kinds of law-abiding actions – marches, peaceful protests. I've changed my lifestyle. Nothing's changed. Or rather, things have changed. They've got much worse.

The last straw was the October 2018 IPCC report, saying we have twelve years to avert climate catastrophe, talking of imminent threats. Our children's future is in jeopardy. When Extinction Rebellion came along, I had this feeling of relief, realizing that this was the last course of action left. I wasn't sure whether non-violent direct action would have any impact, but it seemed the only thing I could do.

I'd trained as a legal observer for various social justice and climate change protests, so I had some inkling of the issues around getting arrested. As a middle-aged white woman living in the UK I felt no reason to fear for my life in this process – in fact I feel a duty to use this privilege to act on behalf of those who can't, for whom arrest could be life-changing or even deadly.

I didn't worry too much about getting a criminal record; I work freelance with no plans to get a salaried job or live in another country, where such a thing might be a problem. I'm not a young person. I can spare a week or longer in prison and it's a sacrifice worth making. I just made sure to pack a good book to read.

So I took part in the February 2019 XR protest at an oil-industry conference, supergluing myself to the entrance of the hosting Intercontinental Hotel. It was a positive experience and felt like the right thing to do. I went through the process: arrest, police station, bail, appearance at court. The whole thing felt human: the police were civilized, professional, polite. As recommended, I called an experienced solicitor, Bindmans (turning down the police offer of a duty solicitor, who might not have been so experienced with laws around protest); I called my husband.

Then the cell. It's a grim little room, with tiled floor and walls, bare except for a ceiling stencilled with anti-drugs message and an advert for Crime Stoppers. There's a wipe-down cushion on a hard bench. The big metal door clangs shut, key scrapes in the lock, bolt is slid home. I'm alone. Occasionally, the odd shout from adjoining cells breaks the silence. The door opens. Release. My solicitor is waiting: friendly and helpful. I was bailed.

The second time, after the DLR action, I went through the same process. Then, at the court hearing, I was denied bail and put on remand. The shock was physical. I was terrified.

Handcuffed, I was taken from court and transferred to a

THIS IS AN
EMERGENCY

high-security prison. The transfer was dehumanizing: in a police van consisting of metal boxes two-foot square. Once in prison, I felt reasonably safe. The prison staff were civil, professional. Among fellow prisoners I tried to keep my body language gently positive. We were in such close contact that, for the most part, talk with the other women in prison came easily; we swapped stories. Many were supportive of what I was doing; others just found it hilarious. Whatever the case, they all got it. Many are mothers themselves; others young women who might become so.

The loss of liberty left me in unbearable physical pain. Cut off from the outside world and my loved ones, I worried over little things, mundane, everyday stuff. I wrote a to-do list for my family and started a diary. Reading and meditation helped, but days felt like weeks.

My stretch in prison became an opportunity to practice relinquishment – a theme discussed in Jem Bendell's paper 'Deep Adaptation' on civilizational collapse that had in part led me here. I was detached not only from my familiar environment and comforts, but also from reasonable considerations: of what will happen today, tomorrow or further ahead; of being there for my children and knowing they're safe. I had no idea how long I would remain in prison: it could have been a two-year sentence. To sit with extreme feelings of anxiety, panic, anguish and fear felt like a training ground for the life we may all have to cope with in future.

There is little privacy. You're locked up from 5 pm to 8 am and also at times during the day. You have to choose between a shower or a phone-call. The bar of soap issued for showers was also to wash my dining plate. You learn to improvise, barter or do without. The prison library wasn't bad. The food was basic, though there was a vegan option. From my cell window I could hear birdsong, see the sky, look again and again at the view over the adjoining association area. Slowly, the shock started to wear off.

I was issued a TV and saw myself in a news montage. It was amazing, watching the media debate change as the week went on. One of Extinction Rebellion's most important functions is to help us with overwhelming feelings of despair at our situation. It doesn't mean everything will work out, but somehow the experience of being with XR, of resisting, helps: we're doing the best we can. Even prison came to feel meaningful – a life-changing experience.

22/ POLICE, ARREST AND SUPPORT

LEGAL TEAM
Extinction Rebellion

Extinction Rebellion is clear that the police continue to be structurally racist, unjust and violent, particularly towards oppressed groups. We are totally opposed to such discriminatory practices. We are taking great care to design our strategy in a way that minimizes the chance that the police will shut down our actions before these actions have reached a critical mass.

Our approach to the police, like that of other non-violent mass movements, is based on what will enable radical change. A key objective is to enable both younger and older people to participate in protests, something which becomes more likely if the police do not overreact. Additionally, the more we can encourage the police to allow our protests to continue through ongoing communication, the more likely it is we can reach that critical mass.

What's more, mass participation often results in loyalty shifts in the police and the army. According to political scientist Erica Chenoweth, social movements that win defections from security forces are nearly 60 per cent more likely to succeed. Approach security forces with determination and compassion in mind: offer them flowers and speak of the joint efforts needed to protect life on this planet. Ask for their views on the key issues. If you can succeed in helping the police and army to question the government orders they are following, this massively increases your chances of disrupting the political system.

Thousands support the rebellion with no intention of getting arrested: their work and support is essential. A few hundred people have decided to take actions which risk arrest and imprisonment, because this is an emergency now: following the laws of a system which is causing death and destruction across the world doesn't always make sense any more.

It is important to train people about what to do if arrested, how to remain non-violent, how not to give information about the movement or individuals to the police when detained. Why never to accept a caution. Why to prepare a statement of conscience for the police interview or why to say, 'No comment, no comment, no comment.'

Remember to support those who are arrested. Turning up outside the police station and waiting through the night to greet people with love, food and music. With a bus fare home, or a place to stay, and to track and follow any cases that are taken to court.

23/ REINFORCEMENTS AND MIDNIGHT SNACKS

WILLIAM SKEAPING
Extinction Rebellion

It's late, very late, and you've spent the entire day outside in Parliament Square. Earlier, the sun was warm, but with darkness the temperature plummeted: jackets and extra layers and heavily sugared tea haven't prevented the goosebumps or stopped your corner of a road-blocking banner from sagging a bit.

The police are tired, too: their faces set and their approach more heavy-handed; sloppiness is creeping in. After hours of standing around, they're wearily dragging a lady in her late sixties out of a tent, four officers carting her off as others try to dismantle a folding gazebo where the biscuit tin was kept. A guy sporting 'Act Now' and Extinction Rebellion badges is on the phone, staring at four shrinking roadblocks, wondering if they can all be maintained. You *know* there's a climate and ecological emergency, you're holding a fifteen-foot sign that says so, but at this point your small part of it feels a bit of a lost cause.

Then, out of nowhere, drums. The sound gets louder. That's not just a few, that's *a lot* of drums. There's an army heading this way. As they come into view, a smile breaks across your face. The goosebumps are back, but this time it's excitement. The reinforcements are here. The energy returns.

A hundred and fifty fresh protesters led by a samba group

do a lap of the block to cheers from all corners. New faces and friends, new supplies of sugar and carbohydrates. The site will be held for another night. The police shuffle back to their positions.

This April, every site of rebellion in central London has had its stories of epic reinforcement. A call goes out for support and appears to be immediately answered. It can feel like a surreal dream or a scene from a movie. Unstoppably large groups of cyclists riding as part of a critical mass are especially empowering: they tour the city fast, bringing energy and news from one site to another, arriving like cavalry, swooping in to cheers.

Such moments of reinforcement don't require huge numbers, just fresh physicality and encouragement in otherwise static sites: signs that others care and are supporting the action. I remember a man on his own walking straight down the middle of a main road holding what looked from a distance like a tartan-wrapped rugby ball under his arm. It wasn't clear if he was with Extinction Rebellion or why he was walking towards us until the sound of bagpipes started. A huge cheer went up: it was as though the sound system had re-started and the area felt busy again.

These aren't happy accidents. Different sites request help with at least half an hour's notice; roving support teams moving between sites help report potential issues in advance and coordinate reinforcements. Make sure that all reinforcements are heavily armed with chocolate bourbon biscuits – I personally believe these to be the true power source of a successful rebellion.

YOU NEVER CHANGE THINGS BY FIGHTING THE EXISTING REALITY. TO CHANGE SOMETHING, BUILD A NEW MODEL THAT MAKES THE EXISTING MODEL OBSOLETE

— Buckminster Fuller

24/ A POLITICAL VIEW

CAROLINE LUCAS MP

Climate breakdown is inseparable from politics. The melting ice caps, the scorching heatwaves and the staggering declines in animal and insect populations are the direct result of failures by people in power.

Irreversible changes to the natural world are taking place because our economy is built on the assumption that precious minerals, fresh air, clean water and rare species can magically regenerate themselves in an instant. That somehow the Earth will expand to meet our insatiable appetites. But the reality is we've stretched our planet beyond its limits – and without a bold reimagining of our economy and the power structures that sustain the status quo, it won't be able to spring back into shape.

Across the world, our political processes have systematically failed to deliver for either people or planet. As Earth's natural systems collapse around us, our society, too, is at breaking point, with spiralling inequality and epidemic levels of mental-health problems.

Faced with these crises, our existing political systems cannot cope, let alone implement the positive solutions we desperately need.

In 2016 people in the UK and the United States gave their establishments a well-deserved kicking at the ballot box. Since then, not much has changed for the better. President Trump is playing a dangerous game with the truth and the law – every tweet he sends another blow to the carefully constructed (albeit imperfect) checks and balances of the US Constitution. And Brexit has brought British politics, and its government, to a standstill.

But even before this current turmoil governments were not taking the climate crisis seriously. For the last ten years, their policy should have been premised on one simple fact: that the vast majority of existing fossil fuels need to stay in the ground if we're to avoid catastrophic climate change. But too often science and common sense have been abandoned in favour of short-term thinking and favours for climate-wrecking corporations. For most in Westminster, and perhaps in many parliaments across the world, protecting the natural world is an optional extra – a nice thing you do on the side to project a caring image to your electorate.

In the UK that complacent, short-termist attitude has led ministers to ditch almost all policies designed to reduce our climate emissions. Since the Climate Change Act was passed in 2008, Conservative-led governments have sold off the Green Investment Bank, a huge publicly owned investor in clean energy. They scrapped nationwide insulation schemes and standards for warmer homes. They prevented the cheapest renewables from competing in the energy market. And they continue to hand out billions in subsidies to the fossil-fuel industry every year.

On the rare occasions she is asked to account for these dangerous policies in parliament, the Prime Minister says the UK is leading the way on climate change. Our home-grown emissions have fallen by 43 per cent since 1990, and that is progress, but it belies the fact that our consumption emissions – the impact of all the goods the UK imports from overseas – have dipped only slightly over the last two decades. We still have so much to do.

If we are to have any serious chance of implementing the radical policies needed to tackle the climate crisis, we must apply the same boldness and urgency to transforming our political system.

In the UK, our parliament's upper chamber consists of 785 unelected members – and our winner-takes-all voting system means even our elected representatives don't reflect the views of the public. At the 2017 general election, 68 per cent of votes didn't get translated into seats, denying millions of people a voice. In 2015, more than a million people voted for the Green Party, but I was the only Green MP elected. Under a proportional system, I'd have twenty-four Green colleagues by my side, helping to expand environmental protections and transform our economy. Electoral reform could revolutionize our politics.

It's time we modernized our electorate, too. Young people around the world taking part in inspirational school strikes over climate change are right to demand that the voting age be lowered to sixteen. My generation's failure to tackle climate breakdown and ecological collapse is unforgivable. Young people will live longest with the consequences of decisions made now, so it's only right that they should have a say in shaping them.

In fact, everyone deserves more of a say over the future of our countries and our world. Extinction Rebellion are right to call for a citizens' assembly on the climate crisis – bringing together representative groups of people to hear objective evidence and make recommendations to politicians. There is so much that needs to change on the regional and local level – so why not find ways for people in towns and cities across the world to play an active part in tackling the climate crisis?

We no longer need to dream up creative solutions to the

ecological problems we face. We have the technology right in front of us. Whatever our political system, now is the time to make decisions and put policies into practice. And this time of crisis presents an opportunity for everyone who believes in building a fairer, greener society. It is now up to progressives of all stripes – from the labour movement to liberal democrats, from trade unionists to environmental campaigners – to seize this moment and campaign for something better.

We need programmes to deliver massive investment in clean energy and affordable public transport, to insulate every home and to bring hope and meaningful work to communities hollowed out by deindustrialization. This Green New Deal can be paid for by measures like wealth taxes, and some initiatives will pay for themselves through the increased tax returns of those in work. Green quantitative easing is also likely to have a role to play, with banks investing directly in green projects, rather than the government handing cash to the banks.

Meanwhile, global leaders need to recognize that we live on a planet with finite resources. Investment on this scale will allow us to explore a new economic model, designed to improve life for everyone while protecting the natural environment we depend on, and measuring our success by people's well-being, instead of company profits.

This is about a paradigm shift in the way the world is structured and the way we live our lives. As Greta Thunberg told world leaders in December: 'If solutions within the system are so impossible to find, maybe we should change the system itself.'

We need to harness the energy in people's anger with the status quo into a movement that gathers in classrooms, living rooms, churches and village halls across the country. A movement that no government and no political party can stop – to redesign our economy so that it gives everyone a role in re-vitalizing our society and tackling climate breakdown.

The crises we face will not be solved by one country alone, but by leaders of all nations speaking out together for urgent and radical change. It's time for politicians to stop arguing among themselves, stop blaming their opponents and unite behind the need for transformative change.

Extinction Rebellion thinks beyond politics. The voices of two politicians are included here as they represent important forward-thinking sentiments that we hope will resonate and develop rapidly within the UK political sphere.

25/ A NEW ECONOMICS

KATE RAWORTH

Calling all economic rebels: humanity's future depends on you. Yes, really. Because, unless we transform the economic mindset at the heart of education, politics, business and public debate, we stand very little chance indeed of thriving in this century.

Take the word 'economics' back to its ancient Greek roots. It means 'the art of household management'. Bring that into our contemporary context and it's the grandest of ambitions: to help manage our planetary home in the interests of all its inhabitants. Doing so calls for no less than economic transformation. Why? Because we have inherited economies that are degenerative, divisive and addicted to growth. And we urgently need to transform them into economies that are regenerative, distributive and able to thrive beyond growth. It's no less than the redesign challenge of our times.

Thanks to the linear industrial system that dominated the twentieth century, today's economies are degenerative by default. We take Earth's materials, make them into stuff we want, use it for a while (often only once) and then throw it away. Take-make-use-lose: it's a one-way process that cuts against the cycles of life by drawing endlessly on Earth's sources – from fish and forests to fossil fuels – while continuously spewing waste into her sinks – from rivers and oceans to the atmosphere. In the process, it destabilizes

Degenerative economy

the life-supporting systems of this delicately balanced living planet on which all of our lives depend.

This degenerative industrial system must now be transformed into a regenerative one: an economy powered by renewable energy in which resources are never used up but are used again and again so that waste from one process becomes food for the next. Nature, of course, has been practising this art for several billion years, decomposing plants and animals into life's molecular building blocks, then building them back up again and again. Now human-made materials – from metals to textiles and plastics – must mimic this art, with open and modular design that enables everything to be restored, repaired, reused, refurbished and (as a last resort) recycled.

This shift from degenerative to regenerative design goes hand in hand with an equally ambitious shift from divisive

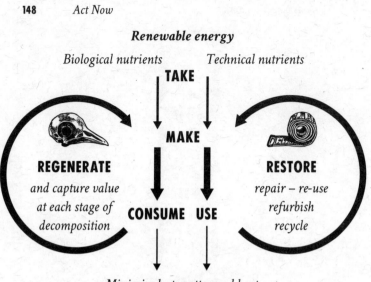

Renewable energy

Biological nutrients   *Technical nutrients*

TAKE

MAKE

REGENERATE
*and capture value
at each stage of
decomposition*

RESTORE
*repair – re-use
refurbish
recycle*

CONSUME USE

Minimise lost matter and heat

Regenerative economy

to distributive economies. Today's deeply inequitable economies are dominated by income and wealth dynamics that tend to concentrate resources and economic power in few hands, resulting in the richest 1 per cent of people globally owning half of the world's wealth. But such extreme inequality is neither inevitable nor immutable. It is an economic design failure.

The good news is that there are unprecedented opportunities to redesign economies to be far more distributive of value created among all those who help to create it. To appreciate the possibilities, go beyond last century's focus on redistributing incomes and embrace the twenty-first-century opportunity of pre-distributing the sources of wealth creation.

For the first time in human history, innovations in four key technologies – how we generate energy, how we make

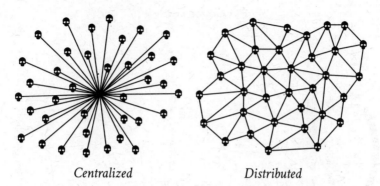

Centralized   *Distributed*

things, how we communicate and how we share knowledge –
are giving us the chance to create economies that are far more
distributive by design. Energy generation is switching from
oil rig to solar panel. Making things is shifting from mega factory to desktop 3D printing. Communications have already
leapt from switchboards to mobile networks. And knowledge sharing is evolving from patents to open source. Of
course, these technologies – especially digital networks – can
also be captured in few hands, but if regulated, designed and
owned for the common good, their distributive potential is
profound. Getting this right is a key twenty-first-century
economic issue.

Now add to this a third redesign challenge – and it's a big
one. Today we have economies that need to grow, whether
or not they make us thrive. What we need are economies that
make us thrive, whether or not they grow. Yes, that small flip
in words hides a major shift in mindset, but it is one we have
to confront. Our economies have come to expect, demand and
depend upon growth never ending. In other words, they are
financially, politically and socially addicted to endless growth.

Today's economies are financially addicted to growth
because at the heart of the current financial system lies the

The Economist's Growth Curve

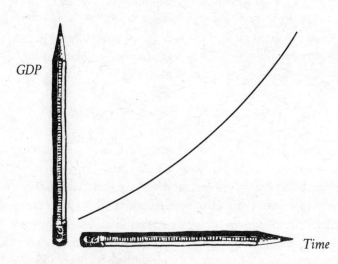

GDP

Time

pursuit of the highest rate of financial return. That in turn puts publicly traded companies under pressure every quarter to show that they have growing sales, growing profits and growing market share. And, since banks create money as debt which bears interest (every time they make a mortgage or a loan), repaying those debts likewise adds to the financial pressure for continual growth.

Our economies are politically addicted to growth because pension funds and the job market have become structurally dependent upon it. What's more, no government wants to lose their place in the G20 family photo, but if their economy stopped growing while the rest kept going, then they would likely be booted out by the next emerging powerhouse. And for governments seeking to raise tax revenues without raising the tax rate, more growth is the all-too-convenient answer.

Our economies are also socially addicted to growth, thanks

to a century of consumerist propaganda, invented by Edward Bernays, nephew of Sigmund Freud. He realized that he could turn the insights of his uncle's psychotherapy into very lucrative retail therapy by convincing the public that the promise of feeling admired, envied, loved or secure is only always just one purchase away.

Here's the real trouble with this structural lock-in to endless growth. Look to nature, which has been thriving for 3.8 billion years, so it's a pretty good example to learn from if we, too, want to stick around. Unlike the economist's growth curve, nature's growth curve gradually levels off to a plateau. In nature, growth is indeed a wonderful healthy phase of life – which is why we love to see children and flowers grow – but in nature nothing grows for ever. Anything that tries to do so destroys itself or the system on which it depends. In our own bodies we know that as cancer.

Nature's Growth Curve

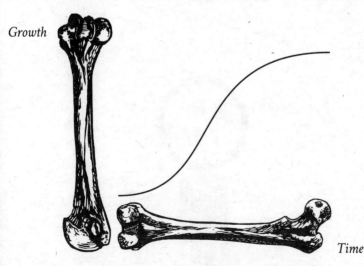

Growth

Time

So what would it take to end this structural dependency on growth and instead to follow nature's advice to create thriving economies that grow until they are grown up? This is the existential economic question of our times and we have barely begun trying to answer it.

For starters, if Bernays were alive today, we'd have to tell him, 'Well done, Edward: the propaganda worked. So now please join the other team and help to undo the insatiable consumerism you helped to stir up.' But it will take far more to end this addiction to endless growth. We need to transform finance so that, rather than being self-serving, it is in service to life. And we need a new basis for relationships between nations so that, instead of competing over GDP, they start to collaborate for collective well-being. These are certainly formidable challenges, but while nations are still fixated on a structural dependency on endless growth, they will not pursue regenerative and distributive economic design on anything like the scale required.

This transformation also calls for a deep shift in our own identity: a rediscovery of who we actually are. Because twentieth-century economics, with its insistent focus on markets, reduced humanity's role in the economy to little more than labour and consumer, creditor or debtor. If the resulting character, widely known as 'rational economic man', were actually depicted in the economic textbooks, he'd have to be a man, standing alone, with money in his hand, ego in his heart, a calculator in his head and nature at his feet.

A narrow portrait for sure, but a dangerous one, too: research shows that the more students learn of this character throughout their studies, the more they come to value competition over collaboration, self-interest over altruism. The lesson? Who we tell ourselves we are shapes who we become.

And that's why the deepest shift required may simply be in reimagining who we are. Market-based economics may define us as labour, consumer, creditor and debtor. But in the household we are partner, parent, neighbour and child.

In relation to the state we are resident, taxpayer, voter, protester. And in the commons we are co-creators, sharers, repairers and stewards. Each of us weaves seamlessly in and out of these diverse economic identities every day, drawing on very different sets of values and skills – from cooperation and empathy to altruism and reciprocity – in order to get along and get by. And nurturing these diverse social skills will be crucial for the transformations ahead.

So this is the work of economic rebels: to create thriving, regenerative and distributive economies that can meet the needs of all people within the life-supporting systems of this unique living planet. Now that's what I call a plan for managing the planetary household in the interests of all its inhabitants.

Who's up for joining the team?

26/ A GREEN NEW DEAL

CLIVE LEWIS MP

In November 2018 riots erupted in France. The Gilet Jaunes (Yellow Vests) movement emerged in response to a major hike in the price of road fuel by Emmanuel Macron's supposedly 'centrist' government. It happened under the guise – and perhaps even with the belief – that this was to assist the transition to a low-carbon economy. But in a country that – like so many others – has an inadequate public transport system, Macron's move had the effect of impacting on any French worker dependent on their car, whether for work or to get the kids to school. It also impacted on any French small business needing a van for commercial viability. Drawing on the *gilets jaunes* that all French drivers are required by law to keep in their vehicle as a symbol, the protest movement spread rapidly across the country: an explosion of anger after years of continual pressure on the standard of living.

For all those who wish to see humanity march towards a low-carbon, sustainable alternative path, there is a critical and simple lesson to be drawn from the Gilets Jaunes movement. It is that any attempt to force through a low-carbon transition by placing the burden on those least able to afford it will fail, and fail badly. The Gilets Jaunes grasped a simple truth that the supposedly sophisticated centrists of the Macron government failed to see: the issue of the climate and the environment is an issue of social and economic justice, as well as ecological justice.

It is this truth that US Congresswoman Alexandria Ocasio-Cortez's Green New Deal bill, now being presented for debate in the US Congress, places front and centre. Aside from a fringe of deniers – admittedly well represented by the buffoon currently occupying the White House – all sides will now admit that climate breakdown, and wider environmental collapse, is an issue for all humanity.

Above all, the Green New Deal presents society with a clear demand: transform the globalized financial system that fuels consumption, climate breakdown, economic crises and social injustice. Make finance servant to the economy and the ecosystem.

The technocratic solutions of the centre have utterly failed their purpose. They are too reliant on 'smart' policy fixes, too beholden to free-market ideals and too unwilling to confront the vested interests that bar our path to a low-carbon, liveable future. Centrist fixes and tinkerings with the system have sustained financial globalization while failing society and the climate.

We must return to the values and principles of the historic left: of social justice, and of trust and faith in ordinary people to transform the world and protect nature's bounty. The strikes by school students hammer home the point, as hundreds of thousands of pupils across Europe walk out of their classrooms to protest and demand action on the climate.

Change won't come from above but from below – and change cannot happen unless it addresses the monstrous injustices that are built into our highly destructive, unsustainable, high-carbon economy. An economy in which the wealthiest 10 per cent are responsible for more than half of all greenhouse-gas emissions. One based on an industrialized farming system that destroys forests, decarbonizes and depletes soil, drives biodiversity loss and yet, globally, receives hundreds of billions in taxpayer subsidies. Above all, an economy sustained by a finance sector that fuels endless consumption with seemingly endless credit – all in the name of vast capital gains, regardless of the ecological cost to both this generation and future generations.

So where to begin? Well, any programme for ensuring a liveable planet must first acknowledge our collective failure to confront capitalism's financial and political power and its commitment to exponential increases in rates of consumption, fuelled by carbon. We must credibly challenge this orthodoxy and secure democratic support for a radical transformation of the carbon-fuelling economy in little more than a decade. To do that, we have to develop a political route map of how we get from where we are now to where we need to be – and fast.

What does that route map look like? First, we need to secure resources for large-scale economic transformational change. That can be achieved by a government committed to subordinating markets in money, goods and services to regulatory democracy – in other words, to the interests of the democratic societies and the ecosystems in which they operate. 'Free-market' neoliberal economic policies that detach markets from society's oversight achieve the reverse. They are designed to subject markets to private, not public, democratic authority.

I don't say this from a dogmatic position but rather one of pragmatism. Let's look at the evidence. We know just ninety global corporations have produced nearly two thirds of the greenhouse-gas emissions generated since the dawning of the industrial age. Half of the estimated emissions were produced in just the past twenty-five years – well past the date when governments and corporations became aware of the damage greenhouse gases were doing to our climate. Deregulation, fragmentation and the globalization of markets in money, goods and services is not the answer. Rather, these are the causes of our accelerating crises.

To tackle these powerful forces we must use democratic regulation to bring offshore capital markets back onshore, ensuring that markets regain their role as servants, not masters, to the real economy, and to society. As both Trump and the Leave campaign were quick to grasp, 'Taking back control' has never been so popular. We must now reclaim and recast this public appetite for greater accountability from markets.

Once this is done, it will be possible for the government, in cooperation with the Bank of England and the private sector, to mobilize financial resources to support the most urgent of society's missions: the transformation of the economy away from its dependence on carbon and towards wider sustainability. We know that can be done, because it is what would be done if Britain were facing any grave threat to our human security.

A brief glance at some of the estimated costs shows why such a mobilization of resources is necessary. It's been estimated at least £500 billion of investment in new low-carbon infrastructure is required over the next ten years to transform the UK economy, of which £230 billion will be required for energy efficiency alone. Transport for the North estimates that between £60 billion and £70 billion must be invested in the north of England's antiquated transport network between 2020 and 2050. The good news is that such investment could add almost £100 billion in real terms of economic benefit to the UK, along with 850,000 new jobs. Across the country, small and relatively cheap transport improvement measures would include improving stations and bus services, park-and-ride facilities, through to bus priority and cycle lanes, traffic-calming schemes, car clubs and bike-hire schemes.

But while there are jobs in physical infrastructure, just as important will be the jobs created to advance education, caring, creativity and the arts. Jobs in medicine, science, music, theatre and literature.

Meanwhile, given the timescales, squeamishness about the state taking on the lion's share of big infrastructure, R&D and other critical, time-sensitive projects must be robustly challenged. The message must be clear: for businesses and public alike, climate breakdown represents a grave threat to our security, and to our futures. Tackling this threat will benefit both the private and public spheres. That's because the 'roaring lion' that is the state – to quote economist Mariana Mazzucato – will provide clear, long-term spending plans to the 'timid mice' that make up the private sector.

For the Labour Party, it would also offer a possible answer to an existential threat to its continued existence as an electoral force: namely, how to reconcile the growing fissure between towns and cities; between its fracturing coalition of the socially liberal and the socially conservative – one so painfully highlighted in its agonized approach to Brexit. That's because such an epic mission – namely the substitution of renewable energy for fossil fuels and labour for carbon – will lead to the creation of a multitude of unionized skilled, unskilled, professional, artistic and creative jobs across the entire country. The so-called 'just transition'.

And thanks to the operation of 'the multiplier', fulfilling that mission will mean that public spending and expanded employment will pay for itself, restoring stability to public finances.

We know it can be done, because it has been done before, when, single-handedly, and almost overnight, President Roosevelt dismantled the globalized financial system known as the 'gold standard'. This freed up his administration from the shackles of Wall Street and enabled the US government to mobilize on a massive scale to tackle the ecological crisis that was the 'dust bowl', as well as the crisis of 1930s unemployment. While many wrongs were enacted against black people under President Roosevelt's New Deal, nevertheless his administration played a critical role in preventing the slide into fascism that so blighted continental Europe at that time.

Of course, while there will be similarities, there will also be different, and bigger, challenges.

This time, we must use every fiscal and regulatory mechanism we have to encourage and support the shift to the so-called 'circular economy'. That means cuts in consumption, more recycling and drastic improvements to resource-use efficiency – especially of our leaky, Victorian housing stock.

We must also abandon the fetish of 'growth' – a fetish first devised by neoliberal economists at the OECD and the *Financial Times* in the 1960s. They sought to encourage exponential economic growth to parallel the vast expansion of the globalized finance sector. Instead, our goal must be a decarbonized economy of full employment, based on renewables, recycling technologies, biodiversity, stock replenishment, sustainable and regenerative agricultural practices and other areas necessary for this transformation.

This transformative programme will redefine what it is to be progressive. As the parliamentary debate on Heathrow's expansion demonstrated, the twenty-first-century Left will split down into two camps. In one will be those who understand that our current fossil-fuelled, consumption-based economy is inimical to life on Earth. That there are no jobs on a dead planet. That real international solidarity recognizes that the jobs of workers in one country cannot come at the price of lives in others (whose countries are already on the front line of the climate crisis). That, ultimately, the international fight for climate justice is the fight for economic justice both within and between countries. That by taking on this fight we can not just save the world but reshape it in the process.

In the other camp will be those who embrace vested interest in return for short-term, diminishing security and prosperity.

The choice facing us couldn't be clearer. It is time to break with neoliberalism, not break the planet.

Extinction Rebellion thinks beyond politics. The voices of two politicians are included here as they represent important forward-thinking sentiments that we hope will resonate and develop rapidly within the UK political sphere.

27/ THE ZERO-CARBON CITY

PAUL CHATTERTON

The climate emergency is also a city emergency. We know that the climate crisis isn't a natural phenomenon. It's a human-made problem, and one that is lock-stepped with the growth of our industrial urban world. Most of the world's population will soon be in cities. They are locked into high-energy throughputs, are responsible for about three quarters of global greenhouse gases, have ecological footprints way bigger than their city limits, and are the beating heart of our pro-growth, consumer-saturated way of life.

Fundamentally changing the way we live in – and design – city life makes sense. It's where the big battles against the climate emergency can be waged. And there will be wider rewards. We need to tackle the climate crisis in our cities in ways that also tackle long-standing urban problems – poverty, alienation, segregation, violence, corporate greed and powerlessness. The way cities respond to the climate emergency will determine the very fate of humanity. We literally need to save the city from high-energy, high-emission, high-inequality life. Here are four areas that can be the focus of our attempts. Let's explore them.

First, the overarching challenge is to repurpose and redesign city life, infrastructures and institutions to meet the target set at the UN talks at Paris in 2015 – to hold global temperature rises to no more than 1.5 degrees Celsius of global warming. While there's a real focus on figuring out a roadmap for how to do this at a global and national level, we have almost no idea what this means for cities, and especially urban neighbourhoods. Exactly how much carbon, what changes, by when, by whom and how? We are largely in the dark.

There are already ambitious targets. Cities across the world are rapidly bringing forward their plans to reach zero-carbon targets. Big global cities like New York, Paris and London are pitching at 2050, while in the UK Bristol and Manchester have brought it into the 2030s and Nottingham by 2028. Emissions-reduction pathways are also littered with confusion. Do they refer to net zero and carbon neutral, which opens up easier routes through emissions offsetting? Or are we aiming for genuine zero-emissions planning, where zero means zero, and now, more realistically, net negative? We have to go beyond zero emissions and start to drastically drawdown carbon from the atmosphere.

The big message is to create city carbon transitions based on social justice. The average citizen has to reduce their emissions to the equivalent of almost three tonnes of CO_2 per person by 2030. What's clear is that the majority of city dwellers across the world, mainly in the Global South, already live well below this level. And the big elephant in the room is Scope 3 emissions from consumption and travel outside city boundaries. Bringing them back on to the city balance sheet completely changes the nature and scale of the task and focuses our attention on reversing almost all externally sourced consumer habits. We are not looking at adjustments any more. It's a complete overhaul of what we know as city life and city economies.

This is the zero-carbon city challenge. It requires locking down an ageing, centralized, corporate-controlled and externally dependent energy system unfit for the challenges ahead. From their buildings, leisure, tourist and retail habits, through transport, workplaces, producer and consumer services, to cities' vast non-renewable energy producers and users. Current city energy systems lock citizens into a brown energy commodity that is priced to concentrate wealth and global political power rather than to create a common good that underpins a flourishing life for all. The negative results are all around us: localized pollution, increases in greenhouse gases, fuel poverty and high utility prices.

In their place, we need to unlock a civic-energy revolution of distributed energy networks, local smart grids, municipally owned energy and zero-emissions, community-led developments. This involves new planning ordinances and mass national retrofit programmes to ensure that every single building is zero carbon; municipal energy companies modelled on the German Stadtwerke, which generate from 100 per cent renewable sources and undercut corporate energy giants; and people's energy action to ensure a 100 per cent moratorium on fossil fuels and fracking. A vast transfer of subsidies underpins all this.

Civic innovations, such as those developed by Repower London, are flourishing, especially in Combined Heat and Power (CHP), onshore wind, solar photovoltaics, anaerobic digestion, local smart grids, energy-storage technologies and the new skills that will underpin these. The new civic-energy sector could really mean that the age of the large power plant is replaced with a constellation of distributed but highly connected small and medium zero-emission energy providers. Every home, garden and street becomes a micro power station. The potential is huge. But will action be connected enough and fast enough to avoid the now dangerous effects of staying on the path of high-emissions urbanism? Part of the equation is demand-curtailment, so city life becomes more localized and less energy-bloated. In essence, zero-carbon targets save cities by starting to untangle the energy system that keeps capitalism, and our ceaseless growth paradigm, going. That's the big challenge.

The second big area of action is transport. This is about the urgent task of how and why we need to lock down city car culture. Almost all modern ills can be understood through the rise of the private fossil-fuel-powered automobile: unnecessary road deaths, the global pandemic of urban air pollution, mounting greenhouse-gas emissions, geopolitical wars, the concentration of corporate wealth and mounting consumer debt, depression, status anxiety, obesity, alienated streetscapes, the decline of vibrant public life and the corrosive effects of individualism.

We simply need to lock down city car culture: privatized, corporate-led, fossil-fuel-hungry automobile dependency, and growth-based planning. This is not about merely addressing the technical issues of designing and building the low-hanging fruit of sustainable-mobility options like bike lanes and mass rapid transit, although of course these are essential first steps. We need to unlock car-free, socially just, zero-carbon, common-owned mobility. Getting rid of cars means getting rid of the conditions that makes us need cars. It's a new mobility paradigm based on responding to climate breakdown, social inequality and living well. This is all absolutely achievable. In only just over one hundred years, we have witnessed the rise of only a handful of cars to a number approaching a billion. The story of the car is so brief it can easily be reversed. But it requires action across culture, infrastructure, work, organizations, behaviour, finance, marketing, power and politics. In particular, it requires vast shifts in subsidies to green and affordable travel and planning decisions that prohibit all new car-based activity. How we choose to get around the city is connected to a set of choices about the very future of the city itself. Driverless cars may yield some marginal emissions gains. But cities full of Google and Tesla driverless cars will not stop the descent into alienated street life, status anxiety and debt-fuelled, corporate-controlled consumerism. The driverless-car city is the next step in the great car take-over of the urban world.

Putting ourselves front and centre of the car-free city is one of the most crucial but difficult tasks ahead. We are not stuck in traffic. We are the traffic. Recognizing our own personal implication in car culture and the damage it is doing is personally threatening. So much is invested in it that it is easier to ignore it. For many of us, perhaps the car represents the only crumb of sanity, freedom and control in an otherwise out-of-control world. It might be the only way we can get food, get our kids to school and get to work in a world full of complicated, expensive and seemingly dangerous options. But we need a step change. Only by designing out the car, and the corporate and fossil-fuel webs that support it, can we save the city.

The third area is the Bio City. An urgent lock-down of the destructive ecological tendencies of urban life is required. The air, water and land ecosystems that cities depend upon are being intensely degraded, and resources are being depleted and commodified. There are vast deadzones of alienated urban sprawl and dereliction, retail areas, highways and industry where residents have little connection with the natural systems that underpin human flourishing. Worse, there's a binary division between us humans and the nature out there. This leads to us treating nature as something external rather than as a life-support system we depend upon.

At the same time, we need to unlock a new 'human–city–nature' deal, which is slowly emerging through restorative and regenerative practices in urban nature. A constellation of pioneering innovators and ideas, including rewilding, permaculture, urban agriculture, continuous productive urban landscapes and blue-green infrastructure is driving this. It is underpinned by a broader shift in the relationship between nature and the city away from resource extraction, private profit, linear notions of progress and privatization and towards equality, stewardship, nature-based regeneration and restorative cyclical and interconnected relations. By reducing wealth and social inequalities and the drive for individual profit, humans can also begin to value and reconnect with each other and the natural world. For too long natural systems have been regarded as a replaceable and free resource at the disposal of the human quest for maximizing individual gain in the service of infinitely growing economies.

A more fundamental rethink of urban nature is underway. Two ideas help here. First, 'biophilia' refers to the innate emotional affiliation of human beings to other living organisms and the strong emotional and psychological benefits that are derived through connection to nature. As an urban design approach, this can replicate the experiences of nature in cities in ways that reinforce that connection. Second, 'biomimicry' refers to emulating or mimicking the complex engineering and design principles found in the natural world. Insights into how nature solves problems can be used to tackle significant social challenges such as climate breakdown or air pollution. Practical applications are emerging through hybrid natural and built forms: living walls, rooftop farms, vertical or sky gardens and breathing buildings. Through these, the challenge remains to create a deep reconnection and love for nature and other species. Without this deep reconnection, most people will simply not see the rationale for protecting and regenerating the natural systems that we depend upon. Reversing the trend of industrial urbanism confronts us with a complex and seemingly inevitable social and political history that incorporates colonialism, imperialism, capitalism and advanced technology. Saving the city requires recovering a very different human–natural deal – one that puts nature and our deep connections with it back in the driving seat. The question remains whether action can be fast enough to respond to the urgent threats of climate breakdown, water stresses, pollution and the annihilation of animal species and biodiversity.

The fourth area for action is what I call the common city. In the contemporary urban experience, something is amiss. It is in fact deeply uncommon. The signature characteristic is the significant and growing gap between the haves and the have-nots, a deep and lasting sense that, for vast swathes of the urban population, what is happening is not for them. Civic democracy has become detached, over-bureaucratic, ossified into silo thinking, and public trust in it continues to plummet. Urban economies no longer attempt to distribute wealth and ameliorate income and social polarities. Instead, they largely function to facilitate large capital enterprises so they can extract value from local economies, suck out and concentrate wealth within extra-local corporate supply chains. We need to unlock a city commons.

Civil society is bursting with potential, ideas and skills to respond to the climate crisis and build community resilience. Examples abound of community- and place-making that challenge the uncommon corporate city and show glimpses of novel forms of citizen housing, common ownership, social/solidarity economy, community wealth, citizens forums, civil disobedience, attempts to revive local places, neighbourhoods and high streets, as well as to reclaim land. The

LIFE

OR DEATH

Cleveland Model, Co-operation Jackson, the shack-dwellers movement, renters' unions, housing cooperatives, open-source digital manufacture and crowd-sourced city plans are all showing how to reverse-engineer city communities and democracies to become places of safety and equality. Saving the city requires building a common city, which can put into reverse the pro-growth capitalist city.

So what's the overall place that this points to? A car-free, negative-emissions, commons-based bio city. This needs to be a positive vision of a beautiful place, thriving communities, abundant commons, pleasurable mobility, sufficient energy, a climate-safe future together. But we have to be realistic. We need to acknowledge and unravel the structural conditions that drive urban unsustainability. The challenge, as always, remains to strategically reflect on what this means. How do micro-examples connect and scale without losing their potency? How do we embed social justice in cities as we respond to climate breakdown? How do we ensure that the way we respond is inclusive and representative of diverse voices? What coalitions of actors will move us in these directions?

We need three things. First, we need a new post-growth storyline – a safe and just operating space. Second, we need an ambitious post-carbon target – and to understand what the task is to comply with the Paris 1.5 target. Finally, we need post-capitalist interventions to create common and civic economies that retain community wealth and build climate-safe solutions. To unlock all this we will need huge shifts in subsidies and primary legislation to raise corporation tax, curtail corporate power and shift the burden of tax from income to land. This will need a new and powerful political movement that can take power and radically redistribute it. It will need ambitious coalitions of city actors prepared to move away from traditional ways of working: renegade entrepreneurs, break-away academics, dissident public officials, rebellious citizens. We need to experiment wide and fast, and we need the finances and political support to do so. This is a climate and city emergency. To save the climate and the city, we need to think big, start small, but act now.

28/ WHAT IF ... WE REDUCED CARBON EMISSIONS TO ZERO BY 2025?

HAZEL HEALY

In October 2018 the Intergovernmental Panel on Climate Change (IPCC) issued a stark warning: enact urgent measures to limit global warming within the next twelve years or irrevocably deplete the ecosystems that sustain human life on Earth.

By way of remedy, the IPCC recommends that we reduce carbon dioxide (CO_2) and other greenhouse gas emissions (GHG) to 'net zero' by 2050. The concept of 'net zero' controversially includes so-called 'negative emissions', which presumes the use of technologies that take carbon dioxide from the air and lock it into underground sinks and reservoirs. This trajectory would give us a chance to limit warming to a maximum of 1.5 degrees Celsius: too hot for the coral reefs, due to be 90 per cent wiped out at that temperature, and no defence against the accelerating impacts such as extreme wildfires, heat waves and hurricanes which are already upon us. Yet in December news came that, after a period of levelling out, carbon emissions were set to rise steeply in 2018.

Perhaps a more vertiginous transition would help focus the mind of world leaders: what if we aimed to cut absolute carbon emissions to zero by 2025? No one knows if it's possible – let alone at this rate – but it's instructive to imagine how such a scenario would play out.

Assuming the world pumped out about 42 billion tonnes

(or gigatonnes/Gts) of CO_2 in 2018, we would have to cut out 6Gt every year – roughly equivalent to total US emissions in 2009 – to hit zero by 2025. There are only two ways to go about this: ramping up clean-energy generation via renewable power and simultaneously massively reducing the energy we use.

Under a zero-carbon scenario, the fossil-fuelled energy sector, which generates around 80 per cent of global emissions, would be phased out fast: coal, oil, gas (including fracking) – the lot. Meanwhile, all heating and transport would have to be weaned off carbon-based fuels and on to clean electricity.

With the tight timescale ruling out new nuclear reactors, the entire world's productive effort would be channelled into a logistical feat unknown in human history: building new solar, tidal, wind and hydro generation while modifying electricity grids; and a monumental efficiency drive in homes and businesses. Zero Carbon Britain suggests building 130,000 100-metre-tall wind turbines to power the UK (in an area over twice the size of Wales), while Zero Carbon Australia proposes twelve concentrating solar plants over 2,760km² (an area the size of Kangaroo Island).

By 2025, renewables – currently supplying 10 per cent of global energy consumption – won't come close to meeting current demand. So our zero-carbon scenario requires the global elite (the 20 per cent of global citizens who account for 70 per cent of emissions) to cut the quickest and deepest. Setting aside climate justice concerns, concentrating on US citizens, who average 16.4 tonnes of CO_2 per person, would bring us closer to zero a lot quicker than concentrating on the people of Niger, who clock up under 0.1 tonne per person.

In the rich world in particular, zero carbon would usher in

a period of huge social change. Energy would be stringently rationed, dedicated to survival and essential activities; we'd go to bed early and rise with the sun. Expect massive disruption in the way food is grown, processed and distributed – more turnips and fewer mangoes on the menu in the UK, for starters. Globally, there would be much-reduced private car use, virtually no aviation, haulage or shipping – spelling a dramatic end to material globalization as we know it.

But energy lock-down wouldn't last for ever. Leading energy and climate scientist Kevin Anderson suggests that, after a few turbulent decades, by 2040 renewables may be generating up to 50 per cent of present-day energy use. Reforestation and habitat restoration will steadily suck carbon out of the atmosphere. Regenerative economies powered by sunlight, as Oxford University-based researcher Kate Raworth puts it, will improve well-being for everyone: society will be more localized, healthy and familiar.

But how to enact change on this scale? To avoid a totalitarian, 'eco-fascist' dystopia, a zero-carbon plan delivered at this rate would need to be contingent on total buy-in, perhaps triggered by a sooner-than-expected suite of apocalyptic impacts such as the collapse of pollinating insects, severe typhoons and saltwater flooding.

A global governance system based on subsidiarity (think a more radical and democratized EU) would then allow each country to find their own path to zero emissions with a slew of policies: carbon taxes, tradeable citizen rations – maybe a basic universal income – held together, suggests Andrew Simms from the Rapid Transition Alliance, by 'a social contract built on principles of fairness and equal access'.

If we imagine an acutely concertinaed transition such as the one outlined here, which brings the IPCC's suggested target forward from thirty-one years to just seven, it helps us to see the scale and extent of change that's needed. And as extreme as carbon cold-turkey may sound, it beats our current trajectory to a 3–4°C planet. As Anderson puts it: 'I just want the status quo – the same world, with a stable climate. What's so radical about that?'

29/ THE TIME IS NOW

CARNE ROSS

The creation of a zero-carbon society that minimizes consumption will require a different kind of politics. There are already enough resources and technology for everyone to have a decent life. But they need to be shared better. The current set-up of the economy, and indeed the way we think about 'growth' or 'wealth' or 'progress', needs to change fundamentally. A dramatic transformation is required, and not only of how we use energy. Today, we have a society and economy that is governed for the benefit of the few. How do we get one that is run for the benefit of the many, or indeed everyone? What kind of politics is necessary to bring about the change that's necessary?

I don't believe that this will be achieved simply by electing a different party into government which will then redistribute wealth more fairly or pass more environmentally friendly policies. The odds are stacked against this in the first place, thanks to a biased media and many institutional blocks put in the way by those who benefit from today's dispensation. But even if such a government were elected, it would find it formidably difficult to make the necessary changes. It's almost impossible to institute the kind of radical reform that's necessary from the top down. And in any case, I don't want political change that is imposed on some who may not want it: that's coercive. Not only is such coercion wrong, it inevitably invites a counter-reaction, and we will be stuck in the same tedious and superficial paradigm of confrontational politics of the 'he said, she said' variety. The necessary transformation is not about political parties. It's about a new conception of how we govern our affairs; indeed, how we deal with one another.

We need to respect a basic – and obvious – logic. We will only get a more equal society if there is equal access to power, and if there is equality of power. You cannot get equality when the few are making decisions for the many. The answer, then, is a system where the many govern the many. In other words, where we govern ourselves. There are a number of names you can give this: direct democracy, mass democracy, municipalism or even anarchism. But the basic idea is the same: everyone gets a shout in decisions about things that matter to them.

We are so used to top-down party politics (but so sick of it) that this sounds radical and perhaps utopian. But such systems have worked in the past – and work today, if only we would pay attention to history and, indeed, to what's going on in other countries. In ancient Greece, one of the birthplaces of democracy (which means rule by the people, the *demos*), citizens took it in turns to govern. Taking part in debates and decision-making was seen as a civic duty, not a professional career for those uniquely fluent in the snappy soundbite. Those closest to the ground were seen as the best qualified to make decisions about those circumstances.

In Porto Alegre, a large city in Brazil, budget priorities (which are in fact the most important political decisions, aside from war) were debated by tens of thousands. Over ten years, this 'participatory' process resulted in a much more even distribution of public services like schools, health clinics and sanitation. Although this outcome was reported by the World Bank, it's obvious: if everyone has a fair crack in the decision-making, it's likely that the resulting decisions will be fair. Notably, party politics fizzled out in Porto Alegre. When

BEYOND
POLITICS

the people took decisions for themselves, the need for parties disappeared. Corruption dramatically declined because no longer could the political elites demand bribes and pay-offs for city contracts now decided transparently.

And right now, in north-east Syria, a system of 'bottom up' democracy has been put into place, and it's working. Thanks to the collapse of the Assad regime in that region, the chance was opened up to set up a new way of self-governing. Decisions are taken at the communal or village level. If the decisions are about larger things than the village, representatives are appointed to regional assemblies. But, crucially, they can only confirm policies that are agreed by the communal assemblies. The representatives are recallable. And very deliberate processes are in place to ensure that women are co-chairs of all assemblies and that religious or ethnic minorities (the region is majority Kurdish) are given an equal voice. I've seen this new democracy – which in fact reflects more ancient and enduring principles – in practice. People like it, and it works. As the war in Syria winds down, there will be a chance to assess it more rigorously and, I hope, for others to learn from it. There, people support the government because it is theirs. It's a long time since I've heard anyone in the West applaud their democratic system.

In Syria it was war that opened the opportunity for radical change. In the West, with deeply entrenched hierarchies and institutions that embody inequality and injustice (and a rapaciously damaging form of economics), the necessary revolution is harder. But you don't get revolutionary change by asking for it or by demanding it from the current political system. You have to *practise* it. We need to change our way of politics, and the culture and habits of how we relate to one another. This isn't about a fancy diagram of new institutions but about changing how we do politics, how we deal with others (and disagreements) and indeed how we live as individuals and as a collective. And we can only learn by getting on with it.

So it's a local approach, at least at first. In the city of Barcelona a new political force is putting these kinds of ideas into everyday practice: assemblies of residents who debate and decide the stuff that matters to them, including how to deal with the excessive tourism that is ruining their city, producing direction for policies that the council and the mayor try to implement. Anyone can set up an assembly, for instance for a local school or hospital. The crucial requirement is that *everyone* with a stake is present and given an equal voice: parents, teachers and, yes, students; patients, staff and families. If it's done right – and it isn't easy – then the resulting decisions will carry far more legitimacy than a sweeping top-down, one-size-fits-all government policy, or the views of the local MP. Indeed, that MP will have to pay attention. If these assemblies spread, a new kind of politics will come into life. Regional and, indeed, national assemblies composed of local representatives could then be established to sort out larger-scale decisions. But, remember, these assemblies are not there to tell the 'lower' – in fact, higher – level local bodies what to do, they are there to implement what they've already decided: it's the very opposite of the system we have to today, where the top decides and everyone else has to implement. What's needed is a system where everyone decides and the national or regional level then administers those decisions.

You can call this 'self-government' or 'bottom-up' democracy. I prefer to call it, simply, democracy – *real* democracy. This is doable. Act, don't ask. Learn by doing. Get on with it. Time is short.

AFTERWORD

ROWAN WILLIAMS

It just might work. It is just possible that sustained pressure will bring about a modest change of heart among decision-makers and 'wealth creators' and some serious adjustments *might* be made.

I can hear the sound of people not holding their breath. It isn't only inertia that we have to contend with, unfortunately. It's vested interests, passionate commitment to the goods and privileges we enjoy because of the way in which we – the collectively wealthy of the world – have chosen to use the material that lies around us.

That 'lies around us'? In a way, there's the problem. We have completely and successfully internalized the belief that the world is made up of dead stuff plus active minds and acquisitive wills. We have lost or suppressed any memory of what it is to live in alignment with the rest of the world, to understand how our bodies themselves 'think', how we are fed and sustained by the flow of life and information through matter. Our technological skills, instead of reminding us of our deep involvement in all this, produce ever more seductive ideas of how we could distance ourselves still further from the bodily structure that simply *is* what and where we are.

We both fantasize and worry about artificial intelligence, that ultimate technological triumph which we like to imagine will cut the umbilical cord uniting us with the complex, beautiful, alarming vortex of material energy we're part of. We dream of liberation through AI when we have never learned or have forgotten the *natural* intelligence that belongs to living in the world as it is, the intelligence that we should be able to see all around us in the interaction of living and non-living existence, the pervasive energy that circulates in the universe, what one modern scientific writer (Kitty Ferguson; thank you, Kitty) wonderfully called 'the fire in the equations'.

So it might not work. Many of the contributions to this book face this possibility (probability?) squarely, which is why it is such an uncomfortable book to read. But what makes it more than a jeremiad or an apocalypse is the sheer energy of conviction here, the conviction that change is worthwhile, whether or not it 'succeeds'. This is because the writers of this book have done some joined-up thinking. The climate crisis is not some unfortunate accident but a reality that has been at the very least accelerated and measurably worsened by a set of habits and assumptions that have poisoned us as a human race. We may or may not escape a breakdown. But we can escape the toxicity of the mindset that has brought us here. And in so doing we can recover a humanity that is capable of real resilience.

To put it very directly: it is worth changing our habits of consumption, the default settings for our lifestyle, the various kinds of denial and evasion of bodily reality that suit us, the fantasies of limitless growth and control, simply because there are healthy and unhealthy ways of living in this universe. To go on determinedly playing the trumpet in a string quartet is a recipe for frustration and collapse and conflict. There are ways of learning to live better, to make peace with the world. Learn them anyway: they will limit the disease and destruction; they may even be seeds for a future we can't imagine.

One of the essays here notes that addictive behaviour

doesn't alter overnight. The priority has to be the plain reduction of harm and risk, the fostering of better habits and (I'd want to add) the creation of strategies to salvage relationships, self-respect, even the possibility of joy. Extinction Rebellion recognizes that the climate crisis is a symptom of far wider kinds of malaise and corruption in the human imagination. That's why the creation of panic, with its inevitable accompaniment of self-protection and withdrawal, is useless in addressing the challenge. We need a fresh sense of the delight to be found in human and non-human creation alike, a fresh sense of the importance of living in attunement with who we are and what the world is.

In the Book of Proverbs, in the Hebrew Scriptures, the divine wisdom is described as 'filled with delight' at the entire world which flows from that wisdom. For me as a religious believer, the denial or corruption of that delight is like spitting in the face of the life-giving Word who is to be met in all things and all people. I long for and pray for change, not just because I want to see my children and their children having a planet to live on, a future that will not be marked by a rising spiral of violent conflict over what is left of the world's goods and a downward spiral of disadvantage and deprivation for the most vulnerable. I long and pray for it because here and now we need to recover our health, our balance – the skill of living with and in the neighbourhood that is this world.

'Neighbourhood' is not a bad word to think about in this connection. We're told on the highest authority to love our neighbours as ourselves; and when Jesus Christ was invited to define who counted as a neighbour, he gave a rather surprising answer. The neighbour is the one who gives you life; and you never know where the next gift of life is coming from, so be alert and ready to love the stranger and the one you think is the enemy. But if the neighbour gives life, then the material world we are part of is obviously a 'neighbour' in the most dramatic way, the life-giving cradle of human existence, the source of air, water, food, not to mention beauty and challenge. 'Neighbourhood', 'neighbourliness' – we all understand pretty much what such words mean, and their homely and prosaic character is itself a reminder that finding a new and fuller way of being human is not at all finding a way of being superhuman (or post-human). It is about settling to inhabit where we are and who we are.

A revolution is a turn of the wheel, and the paradox of true revolution is that it takes us back from insanely dangerous places to having our feet on the ground again – coming back to where we started and knowing the place for the first time, as T. S. Eliot puts it in his greatest poem. In this time of massive public denial and displacement – so miserably evident in the ego-boosting dramas of the Brexit debates and the resurgence of surly, self-protective localism across the world's political landscape, Extinction Rebellion urges on us the revolution of coming to ourselves, coming to truthfulness, healing the broken connection with what we are.

It might just work. It might allow a new space and a new imagination to flower in the face of incipient tragedy, a new hope and dignity for human agents, not least among the young, who can so easily feel completely ignored and unvalued in a world apparently indifferent to *their* future. Change the narrative, and who knows what is possible? Accept the diseased imagination of the culture we have created and the death count begins now. Anger, love and joy may sound like odd bedfellows, but these are the seeds of a future that will offer life – not success, but life.

WHAT IS YOUR PLACE IN THESE TIMES?

GAIL BRADBROOK

Standing above the crowds in Oxford Circus, in the pink Extinction Rebellion boat, Daiara Tukano spoke of existence as resistance. Coming from the Tukano indigenous nation of Brazil's Upper Rio Negro, a community enduring severe human rights abuses and under sustained environmental attack, she told us that indigenous nations protect 82 per cent of the Earth's biodiversity. Her message to us was that if you are alive at this moment in history, it is because you are here to do a job.

So what is your place in these times? Have you felt the call to join Extinction Rebellion? Which of your gifts are needed right now? Maybe you feel ill equipped. Bring your uncertainty, together with a willingness to learn. You may feel your gifts are simple. Simple offers, made with true love, are the stuff of life. Do you feel the call to be with us on the streets? Come along and remember the power of togetherness, when the people are determined and strong. Join us in our home communities, let us grow as we are needed. You are so very welcome.

These are times of unravelling, dissolving, transformation. Don't expect to be the same person as before you took part in this journey. For each of us there is an individual challenge, there are waves of difficulties, obstacles, challenges that can be hard to anticipate and hard to name. It's time to trust what is happening and to be willing to be changed.

We have shown in the UK something of what we are made of – which is perhaps fitting for the nation that unleashed this incredible and destructive industrial society on the world. Our challenge now is to look beyond our island nation and see with fresh eyes the rest of our family, spread across the world. To open our hearts. When we are able to fully feel the losses among us, then we will be able to do what these times truly require from us. All the children are our children. We can protect those closest to us only when we remember our love for those furthest away. This is an international rebellion, aligned with all peoples living with struggles to protect life on Earth. This is sacred.

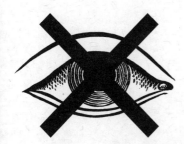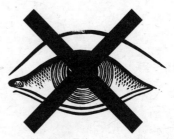

TIME TO
STOP
READING

BLOCK A ROAD

WHAT YOU'LL NEED

- A banner long enough for the width of the road
- Rebels to hold it
- Literature to explain your action (and XR's values)
- A Rebel timekeeper (seven minutes blocking, three minutes off the road, and repeat)
- Cakes for pissed-off drivers (and Rebels to talk to them)
- Flags, badges, stickers and placards
- Music (it's hard for drivers to go ballistic if you're having a disco)
- Wellbeing and legal-observer Rebels

SHUT
A BRIDGE

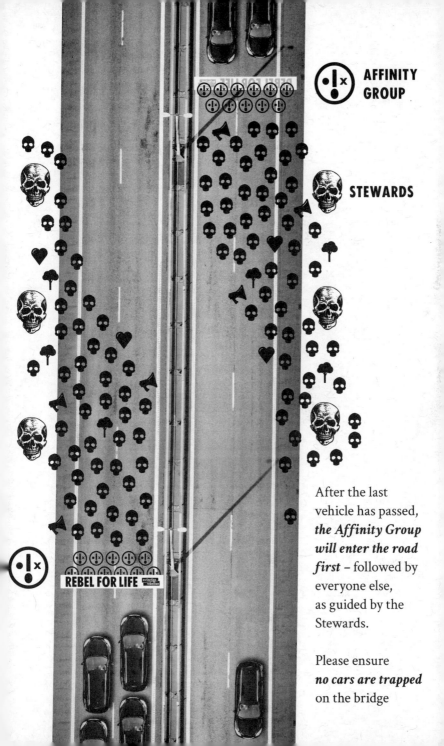

AFFINITY GROUP

STEWARDS

REBEL FOR LIFE

After the last vehicle has passed, *the Affinity Group will enter the road first* – followed by everyone else, as guided by the Stewards.

Please ensure *no cars are trapped* on the bridge

REBEL
FOR
LIFE

WHAT ARE YOU GOING TO
DO WITH IT ?

Acknowledgements.
Cover and page design/artwork: This Ain't Rock'n'Roll – Clive Russell & Charlie Waterhouse.
Artwork: woodblock prints, back cover and pages 3, 15, 18, 48, 49, 50, 89, 103, 120, 123, 124, 133, 148, 150, 151, 178, 187 and 193, Miles Glyn; 52 and 53, Nan Goldin; handcuffs, page 133, Arthur Stovell; 'What are you going to do with the World', p. 195, David Shrigley.
Photos: title page and p. 140, Dee Ramadan; page 39, Adam Hinton for Project Pressure; page 118, Lauren Marina; other photography by This Ain't Rock'n'Roll.

THE SOCIAL CONTRACT

An agreement on: [INSERT DATE]

BETWEEN *1/ The State*
AND *2/ You, the Citizen*

This is a Social Contract between **You, the Citizen**, and **The State**. It binds you. It binds The State. As long as we both keep up our ends of the deal.

PREAMBLE
There are billions of us now. We need, somehow, to live together. Let's cut the grand words and legalese and speak plainly about our duties towards each other. It's too important for anyone to misunderstand.

DURATION OF THE CONTRACT
This will last forever, or until one or both sides breaks the terms.

THE STATE AGREES:
1/ I am **The State.** I agree to look after you. That's the whole point of me.
2/ I agree to protect you. I'll make laws so that everyone knows the rules. I'll need courts, a police force and army to enforce the rules. But I'll only use them for that purpose.
3/ I'll make sure the laws apply to everyone – even to me and the people who work for me. That's called the rule of law.
4/ I'm going to defend your human rights. So you have freedom from threats of violence, torture, slavery, unnecessary imprisonment; even death – if I can prevent

it. And you'll have freedom to live your best life, in privacy, with your family, in a home, in health and welfare, choosing your religion, saying what you want to and protesting when you must. You get all these rights whoever you are - I'm not going to change them depending on what group you're from.

5/ There's no point having rights if you can't breathe. So I'm going to make sure you have clean air and uncontaminated water. I'll stop your home being flooded if I can avoid it. I'll make sure the temperature stays at a safe level. And if things get really bad, I'll take urgent action. After all, I exist to look after you.

YOU, THE CITIZEN, AGREES:

6/ You are the citizen. You don't get **1/** to **5/** for free. You have responsibilities too. The main one is to follow the laws which **The State** sets for you and your fellow citizens. As long as they are understandable and fair.

7/ **The State** needs resources for all the stuff it gives you. So you'll chip in with a bit of what you earn – that's tax.

8/ You have a responsibility towards your fellow citizens. **The State** can't do everything for you. So if something is going wrong, you are going to have to stand up and be counted.

That's it. In summary – The State will hold up its end if You, The Citizen hold up yours.

SIGNATURES
The State   *You, The Citizen*

IF YOU FEEL THAT THE STATE HAS BREACHED THE SOCIAL CONTRACT, RIP OUT THIS PAGE AND JOIN EXTINCTION REBELLION